DAMPING IN FIBER REINFORCED COMPOSITE MATERIALS

Woodhead Publishing Series in Composites Science and Engineering

DAMPING IN FIBER REINFORCED COMPOSITE MATERIALS

PRAMOD KUMAR
Assoc. Professor, Dr. BR Ambedkar National Institute of Technology (NIT) Jalandhar, Punjab, India

S.P. SINGH
Professor, Department of Mechanical Engineering, Indian Institute of Technology Delhi, Hauz Khas, New Delhi, India

SUMIT SHARMA
Assistant Professor, Dr. BR Ambedkar National Institute of Technology (NIT) Jalandhar, Punjab, India

ELSEVIER

WP
WOODHEAD PUBLISHING
An imprint of Elsevier
elsevier.com/books-and-journals

Woodhead Publishing is an imprint of Elsevier
50 Hampshire Street, 5th Floor, Cambridge, MA 02139, United States
The Boulevard, Langford Lane, Kidlington, OX5 1GB, United Kingdom

ISBN: 978-0-323-91186-3

For Information on all Woodhead Publishing publications visit our website at https://www.elsevier.com/books-and-journals

Publisher: Matthew Deans
Acquisitions Editor: Gwen Jones
Editorial Project Manager: Tom Mearns
Production Project Manager: Fizza Fathima
Cover Designer: Miles Hitchen

Typeset by Aptara, New Delhi, India

Contents

Preface

The primary audience targeted for this book include researchers working in the area of damping in composite materials. Currently, no book on the market explains the basics of damping in composites. The proposed book will cover the basics of damping in composites, and modeling (in Matlab), as well as in FEM. The book will also include the basics of the dynamic behavior of composites and will explain the use of a dynamic mechanical analyzer in predicting damping in composites. People working in the field of mechanical engineering, industrial engineering, biotechnology, and physics will find this book useful in predicting the damping behavior of fibrous composites. Furthermore, the book can be used as a textbook on "Damping in Composites" for postgraduate and doctorate-level students. Those who want to learn more about the basics of damping in composites will find this book extremely helpful, as there is no other book published on this topic.

Damping is an important parameter for measuring and predicting the dynamic performance of composite materials. In this exemplary new book, the authors discuss damping behavior in fiber-reinforced composites. Divided into 7 main chapters, the book starts with an introduction to the basic concepts of damping in composite materials. Methods of modeling damping are then discussed in chapter 2. These include both macro and micro-mechanical approaches. Chapter 3 deals with experimental methods for measuring damping. The decay plot and circle methods have been discussed in detail. In chapter 4, a parametric study of a two-phase composite material is presented using different micromechanical models, such as unified micromechanics, and Hashin Eshelby's to predict elastic moduli and loss factors. A bridging model that incorporates the effect of fiber packaging factors is then compared to FEM results. Chapter 5 investigates the effect of the interphase on the mechanical properties of the composite. A mathematical model is developed for the prediction of elastic moduli of a three-phase fiber-reinforced composite. Chapter 6 presents a nonlinear model for the prediction of damping in viscoelastic materials. The final chapter looks at some of the main engineering applications of damping in real life and principles of vibration control. The readers will be able to infer the importance of damping and principles of vibration control in FRPs after reading this chapter.

The book will be an essential reference resource for academic and industrial researchers working in the field of composite materials, especially damping behavior.

Pramod Kumar
Assoc. Professor, Dr. B. R. Ambedkar National Institute of Technology
Jalandhar, Punjab, India
S.P. Singh
Professor, Department of Mechanical Engineering,
Indian Institute of Technology Delhi,
Hauz Khas, New Delhi
India
Sumit Sharma
Assistant Professor, Dr. B. R. Ambedkar National Institute of Technology
Jalandhar, Punjab, India
June 2022

CHAPTER 1

Introduction

A structural composite is a material system consisting of two or more phases on a macroscopic scale whose mechanical performance and properties are designed to be superior to those of the constituent materials acting independently. One of the phases is stiffer and stronger and is called reinforcement and the less stiff and weaker phase is known as matrix. Because of chemical interaction or other processing effects, an additional distinct phase called interphase exists between the reinforcement and matrix. The properties of composite materials depend on the properties of its constituent, their geometry, and the distribution of the phases. The properties of the composite materials combine the best features of each constituent to maximize a given set of properties, that is, stiffness, strength-to-weight ratio, tensile strength, and minimize others, such as weight and cost. Composites have a unique advantage over monolithic materials, such as high specific strength, high specific stiffness, tailored damping, and adaptability to the intended function of the structure. These materials are being used extensively for various high technology applications, such as spacecraft and aircraft structural components, gas turbines, marine, and automobile applications. The endless research for reliable and low-cost structural and material system resulted in inexpensive fabrication methods, which have made composites affordable to several appliances.

POLYMER-MATRIX composites (PMCs) and metal-matrix composites (MMCs) are two of the broad categories of composite materials in terms of matrix classification. PMCs are typically used in low-temperature structural applications, such as in civil structures, biomedical implants, automobiles, and airframe structures. The fibers typically provide the stiffness and strength to the composite and can be made from a wide variety of materials, including glass, graphite, Kevlar, and boron, as examples. The fibers can be arranged in almost any fashion, ranging from totally random to highly structured and organized. In MMCs, the matrix phase consists of continuous metallic material, such as aluminum, titanium, magnesium, copper, etc. The reinforcing constituent is normally ceramic (e.g., silicon carbide, silicon nitride, alumina). MMCs are used in the aerospace industry for airframe and spacecraft structures, as well as in the automotive, electronic, and even leisure

Damping in Fiber Reinforced Composite Materials.
DOI: https://doi.org/10.1016/B978-0-323-91186-3.00004-6

industries. BECAUSE OF THEIR attractive specific stiffness and strength, the study of the mechanical behavior of fiber-reinforced PMCs and particle-reinforced MMCs is of great interest to researchers and engineers in many sciences and engineering disciplines.

Composite materials being inherently heterogeneous are distinct from metals because of various reasons like mismatch in properties of the constituents, characteristic nature of the interface, the interaction of the constituents at the micromechanical level, and their peculiar modes of failure. Hence, composite research is focused on maximizing the potential of mechanical performance through microstructural design and requires a thorough understanding of the micromechanical interaction processes between the matrix and reinforcing materials. Traditional continuum mechanics, based on continuity, isotropy, and homogeneity of solids, is not directly applicable to heterogeneous composites, since, microscopically, fibers and particles are present within composites and have a significant effect on their overall properties. Thus, micromechanics is applied to analyze the relationship between material property performance and material structure on a finer scale (i.e., microscale), which encompasses mechanics related to microstructures of materials. Several analytical and numerical studies in the form of macromechanical, micromechanical, and structural models/theories to predict the static and dynamic performance of the composites are reported in the literature.

Damping is an important parameter pertaining to the dynamic performance of fiber-reinforced composite structures. Damping of vibrations has become an important requirement in the design of automotive and aerospace structures. Design considerations involve minimization of resonance amplitudes and extending the fatigue life of structures subjected to near-resonant vibrations under suddenly applied forces. In particular, polymer composites have generated increased interest in the development of damped structural materials because of their low density and excellent stiffness and damping characteristics. It appears that design changes that cause an increase in damping will cause corresponding reductions in stiffness and strength. It has also been shown that the improvement of damping can be achieved by active and/or passive means. Active damping control requires sensors and actuators, a source of power, and a compensator, which gives good performance under vibratory conditions. Passive damping control consists of the use of structural modifications, damping material, and/or isolation techniques. Passive damping typically requires high loss viscoelastic or fluid materials and thermal control. Material damping can contribute to the passive control system by using the inherent capacity of the material to dissipate vibrational energy. Because of reduced system complexity, passive damping generally contributes more effectively to the improvement of the reliability of machines and structures than does active damping. Also, some passive damping may be required to have a stable active control system.

The successful characterization of the dynamic response of viscoelastically damped composite materials to prescribed modes of loading and time histories depends upon the use of appropriate analytical models/methods. The important considerations are the properties of composite materials based upon their constituents and their interaction, defects, and selection of computational techniques. Damping studies in fiber-reinforced composites at the micromechanical and macromechanical level are mainly carried out with the consideration of material as linearly or nonlinearly viscoelastic. However, most of the work available in the literature for the prediction of damping is for linear viscoelastic composites and makes use of the elasto-viscoelastic correspondence principle [1] or strain energy method [2]. The phenomenon of dissipation of energy in fiber-reinforced composites under cyclic loading is distinct from that in metals.

1.1 Objective of the book

The goal of this book is to develop realistic models for damping exhibited by polymer composites. The nature of damping and its dependence on strain amplitudes, as well as the frequency of excitation is needed to be analyzed in detail. This is intended to be accomplished with the following specific objectives.

i. Study of two-phase damping models and parametric evaluation of the effect of fiber volume fraction, fiber packing factor, and frequency.

ii. Development of three-phase bridging models for prediction of various damping coefficients of polymer matrix composites, including the effect of fiber volume fraction, fiber packing, and frequency.

iii. The role of interphase on the mechanical properties of the composite is well established. The present work is an attempt to further justify this role for the optimum design of coated fibers.

iv. Experiences have indicated some changes in damping capacity with the frequency. In this work, an attempt has been made to look into this effect.

v. The damping ratio in composites also depends on strain amplitudes, especially when the matrix behaves nonlinearly. An attempt is made to model this nonlinearity in damping which can be used for reliable predictions of stress where large deformations are involved.

1.2 Outline of book

This work is organized into 7 chapters with the following contents:

Chapter 1 consists of an introduction of fiber-reinforced composite materials, the scope of the present work, an outline of the dissertation. Chapter 2 presents the literature review of the past work related to damping considering micromechanical and macromechanical approaches. Different damping prediction methods and experimental measurement methods for damping are discussed. Past research work on nonlinear damping is also reported.

Parametric studies in composites involve the determination of damping as a function of its geometric parameters: fiber angle, laminate configuration, thickness, width and length of the specimen, fiber diameter, etc. In Chapter 3, the effect of the above stated as well as other factors, such as stress, strain level, stiffness, frequency, etc., have been considered while determining the damping of composites. Recent developments in measuring techniques for the identification of damping in composite materials (local techniques like polar scanning and global techniques like the resonalyser method (based on measuring modal damping ratios of composite material plates) have been discussed in detail. The imaginary parts of Poisson's ratio and shear moduli have been discussed (DMA reveals only the imaginary part of Young's moduli). Also, practical examples proving that damping is important, have been added. Dynamic mechanical analyser (Tritec 2000) was used for the measurement of loss factors and dependence of frequency and strain amplitude was evaluated for different fiber volume fractions. Zwick universal testing machine (1445 model 10 kN) was used for cyclic loading and loss factors were obtained.

Chapter 4 investigates the effect of fiber packing factor on loss factors for two-phase composite using the two-phase bridging model. A parametric study has been accomplished for different models like the Hashin model, Eshelby's model, and Servanos model. FEM modeling in standard FEM software, NISA 15, has been carried out for hexagonal and square packing geometry and the optimization technique available in MATLAB is used to obtain the fiber-packing factor for the hexagonal and square array. The effect of frequency on loss factors have been modeled and discussed for the different fiber volume fractions of glass fiber-reinforced epoxy (GFRE) composite.

Chapter 5 presents the mathematical modeling for three-phase composite materials considering the effect of the fiber packing factor. The effect of interphase moduli and fiber volume fraction on loss factors has been explained in detail. Dependence of loss factor on frequency has been discussed for three-phase GFRE.

Chapter 6 deals with the mathematical modeling of nonlinear damping. A modified viscoelastic four-parameter model has been used to obtain the stress–strain behavior for cyclic loading. Longitudinal and transverse loss factors have been obtained using a hysteresis loop obtained by a stress–strain plot for cyclic loading. In Chapter 7, the applications of damping in fiber-reinforced composite materials have been discussed in detail with the help of several examples. The principle of vibration control in fiber-reinforced composites has been discussed at length.

References

[1] Hashin, Z. Complex Moduli of Viscoelastic Composites: I General Theory and Application to Particulate Composites, International Journal of Solids and Structures, 6 (1970) 539–552.

[2] Ungar, E.E., Kerwin, Jr. E.M. Loss Factor of Viscoelastic System in Terms of Energy Concepts, Journal of Acoustical Society of America, 34 (1962) 954–958.

CHAPTER 2

Methods of modeling damping

Everything we use is composed of composite materials, from computer chips to flexible concrete skyscrapers, from plastic bags to artificial hips, from optical cables to automobiles. Designing and making these materials innovatively and efficiently impacts all areas of our present and future needs. Fiber-reinforced polymer composites due to their exotic properties, such as high specific strength, high specific stiffness, corrosion-resistance, fatigue-resistance, chemically inert, nonmagnetic, high damping qualities, durability, tailorability, low maintenance, etc. are widely in use. In addition, the viscoelastic character of composites renders them suitable for high-performance structures in aerospace, marine, and automobile applications. However, these materials are quite distinct from metals because they exhibit several peculiar modes of failure (matrix cracking, delaminations, fiber breakage, and interfacial bond failure due to debonding) and interaction at the micromechanical, that is, constituent level. Several analytical approaches in the form of micromechanical, macromechanical, and structural models/theories are available in the literature. Results of investigation carried out by the researchers for both static and dynamic performance of composites are reported.

2.1 Introduction

Damping is one of the most structure-sensitive properties, on both the micro and macroscopic scale, that can be easily measured. As a result, damping measurements provide one of the more sensitive inspection tools. Damping measurement is also used as a macrostructural research tool for studying stress, strain, and slip at interfaces within structural configuration and systems. Damping in fiber-reinforced composite materials can be incorporated as desirable by components. The successful characterization of the dynamic response of viscoelastically damped composite materials to prescribed modes of loading and time histories depends upon the use of an appropriate analytical model/method describing properties of composites based upon

Damping in Fiber Reinforced Composite Materials.
DOI: https://doi.org/10.1016/B978-0-323-91186-3.00003-4

its constituents and their interaction; condition of interphase and presence of defects and selection of computational techniques. Composites are anisotropic and nonuniform bodies, and a description of damping processes in these materials calls for essential new developments in the theory of damping. Zioniev and Ermkov [1] have classified the studies on composites as experimental investigations to generate damping data, the value, and meaning of damping characteristics, their relation with material internal structure, and development of damping theories/models to fully describe energy dissipation processes in composite materials. A critical review of the available literature on composite material damping has been attempted here with regard to different aspects, such as mechanisms of damping, methods for predicting damping, and damping models/theories.

2.1.1 Different damping mechanisms for composite materials

There are many sources of energy dissipation in fiber-reinforced composite materials as discussed below:

(a) Due to the viscoelastic nature of its constituents (matrix, fiber, and interphase): Matrix materials are always viscoelastic and fibers like Kevlar and carbon also have damping. Interphase is the region adjacent to the fiber surface all along the fiber length. The nature of interphase may be weak or strong and that may also affect the evaluation of damping of composite materials.

(b) Damping of composite is increased due to damage like a crack in matrix, broken fiber, or due to slip between fiber and matrix interface [2]. As discussed by Chandra [3], damping is more sensitive than stiffness due to damage in composite materials.

(c) Damping due to viscoplasticity: Fiber-reinforced composite materials exhibit nonlinear damping due to the presence of high stress and strain at large amplitudes of vibration. Thus, consideration of elastoplastic micromechanical behavior is important even for applied stresses that are intended to remain below the apparent elastic limits of composite materials.

(d) Due to thermoelasticity: In the case of cyclic loading materials get heated and energy is dissipated. The energy dissipated per cycle is related to the rise in temperature [4] for thermoplastics. Thermoplastic composites show a high temperature rise, which is a function of applied load, frequency, sample thickness, and the number of cycles [5].

2.1.2 Methods for damping prediction

Prediction of damping at the micromechanical, macromechanical, and structural level based on the assumption of linear viscoelasticity are carried out by many analytical models. Elasticity approaches and mechanics of material have been utilized for elastic solution of moduli, and the damping is further predicted using two different methods as underlined below:

(a) *Correspondence principle*: The correspondence principle [6] states that the linear elastostatic analysis can be converted to a dynamic linear viscoelastic analysis by replacing static stresses and strains with corresponding dynamic stresses and strains, and by replacing elastic moduli or compliances with complex moduli or compliances respectively. This method has been applied to micromechanical models to predict damping in aligned discontinuous and continuous fiber-reinforced composites [7,8] The correspondence principle has also been used in combination with classical lamination theory (CLT) to determine loss factors for laminated composites [9]. The loss factor has been expressed as the ratio of the imaginary extensional stiffness to the real extensional stiffness. Hashin [6] successfully applied the correspondence principle developed for isotropic viscoelastic materials to anisotropic fiber-reinforced composites to predict complex moduli. Thereafter, this principle found wider acceptability in the work of several researchers [10–12]. Some general assumptions, such as fibers are elastic and nondissipative, the matrix is elastic in dilation but viscoelastic in shear are made in the above works.

(b) *Strain energy method*: This method relates the total damping in the material or structure to the damping of each element and the fraction of the total strain energy stored in that element. It states that for any system of linear viscoelastic elements, the loss factor can be expressed as the ratio of summation of the product of the individual element loss factor and strain energy stored in each element to the total strain energy.

2.1.3 Methods for measurement of damping

(a) *Log decrement method*: Measure of damping can be made from the transient response of the system in Fig. 2.1. The log decrement method computes damping from the rate of decay of the system response in the time domain. Computing the damping by this method is illustrated in

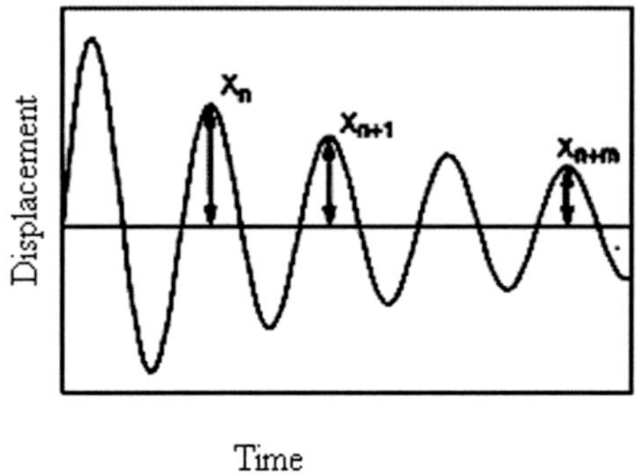

Figure 2.1 Log decrement method.

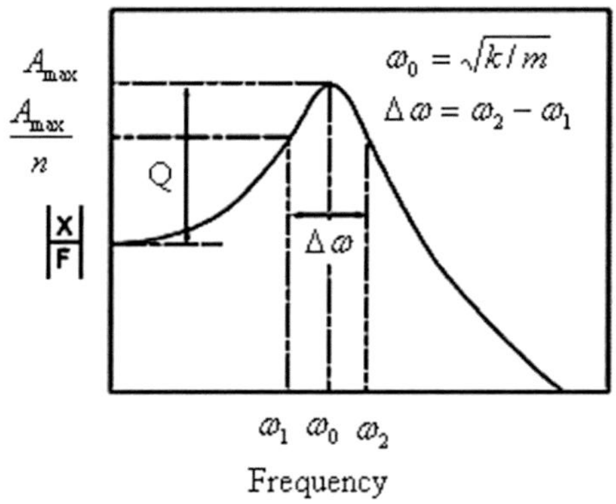

Figure 2.2 Half power bandwidth method.

Fig. 2.3 and is defined in Eq. (2.1).

$$\delta = \frac{1}{m}\ln\frac{x_n}{x_{n+m}} \tag{2.1}$$

(b) *Amplification factor method*: Measurement of damping for the SDOF system of Fig. 2.1 is the amplification factor Q which is the ratio of the response amplitude at resonance ω_o to the static response at $\omega = 0$. The amplification factor relates to the hysteretic loss factor η through Eq. (2.2).

$$Q = \sqrt{(1+\eta^2)/\eta} \tag{2.2}$$

(c) *Half-power bandwidth method*: For the SDOF structure defined by Eq. (2.3) the structure will possess a classic compliance response as shown in Fig. 2.2. The level of damping can be subjectively determined

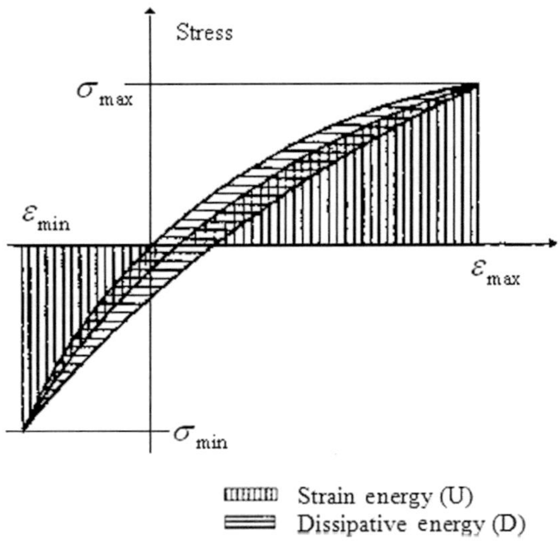

Figure 2.3 Hysteretic loop for damping measurement.

by noting the sharpness of the resonant peak at $\omega_{o:}$ the more rounded the shape, the more damping present in the structure. For a quantitative measure of damping, the half-power bandwidth method can be employed. As defined in Eq. (2.4) the damping of the structure $\boldsymbol{\eta}$ can be determined from the ratio of $\Delta\omega$ to ω_o with $\Delta\omega$ determined from the half-power point down from the resonant peak value, $A_{\max}$ (equal to the inverse of the amplification factor Q). On a decibel scale, this corresponds to a 3 dB drop from the peak. For that reason, this damping measurement technique is also referred to as the 3 dB method.

$$m\ddot{x}(t) + \overset{*}{k}x(t) = F(t) \tag{2.3}$$

where, $F(t) = F_0 e^{i\omega t}$, $\overset{*}{k} = k(1 + \eta\, i)$

$$\eta = c\frac{\Delta\omega}{\omega_0} \tag{2.4}$$

where, $c = \frac{1}{\sqrt{n^2-1}}$ and $n = \sqrt{2}$ for half power bandwidth.

(d) *Hysteresis loop method* [13]: Damping measurement has been carried out by calculating the energy loss per cycle of oscillation due to cyclic loading. By plotting the instantaneous stress versus strain for a given cycle of motion, the hysteresis curve of Fig. 2.3 is generated. The area captured within the hysteresis loop D is equal to the dissipated energy per cycle. The loop area is used to calculate damping, as shown in Eq. (2.5).

$$\text{Damping} = \frac{D}{U} \tag{2.5}$$

$$\text{Loss factor } (\eta) = \text{Damping}/2\pi \tag{2.5a}$$

$$\text{Dynamic Modulus} = \frac{\sigma_{\max} - \sigma_{\min}}{\varepsilon_{\max} - \varepsilon_{\min}} \tag{2.6}$$

When applying these methods to composites, the composite becomes a system, and the nature of the element depends on whether the analysis is micromechanical or macromechanical. In micromechanical analysis, the elements include the constituents, such as fibers, matrix, and their interaction, void content, and the interphase. On the other hand for macromechanical analysis, the individual lamina is the element whose strain and dissipation energies combine to give the overall loss factor of the laminate.

The finite element method and certain other numerical methods for optimization techniques are being applied either for the analysis of the damping of composite materials or optimizing the damping of composite structures. Each of these approaches and methods has its scope and limitations concerning the damping prediction in composites. The progress of research activity in the various directions as related to damping in fiber-reinforced composite materials has been reviewed in detail as under.

2.2 Macro mechanical approach

Prediction of damping of laminates by considering the properties of laminates is known as the macromechanical approach.

2.2.1 Studies on laminate damping

Damping measurement was pioneered by Adams et al. [14,15] experimentally for the characterization of composite materials and measured quantities are used for the development of analytical models. Adams and Bacon [15] developed a macro mechanical model for damping in unidirectional fiber-reinforced composites, which is now being referred to as *Adams-Bacon criteria*. Adams-Bacon damping criterion has been validated for its accuracy by several investigators [16,17]. It states that the energy dissipated in a thin unidirectional lamina is the sum of the separate energy dissipated due to longitudinal stress, transverse stress, and shear stress. Consequently, the specific damping capacity can be defined as the ratio of energy dissipated to strain energy stored. McIntyre and Woodhouse [18] studied the dynamic behavior of orthotropic sheet materials within the approximation of thin plate bending theory, underlying that the linear vibrational properties are governed by four elastic and four damping constants corresponding to any frequency. Experimental verification of damping and frequencies showed good agreement with predicted values. An analytical macromechanical model based on the elastic-viscoelastic correspondence principle has been developed to predict composite loss factors incorporating its frequency dependence [19]. Using classical laminate theory, complex A-B-D matrices have been obtained and longitudinal, transverse and shear loss factors are determined. The loss factors thus predicted show that inconsistencies documented in the literature on fiber orientation at which a maximum loss factor occurs can be resolved by incorporating the frequency dependence of the complex loss factor. This occurs because, for a given investigation, the test specimen dimensions are typically kept constant. When the fiber orientation of the test specimen is changed the stiffness also changes, resulting in the change in the first resonant frequency of the composite beam. The model shows that the stress coupling due to unbalanced laminated constructions has a pronounced effect on the loss factor. The loss factor increases proportionally to the magnitude of the bending twisting coupling terms. Thus, loss factor increases much more for orientation of 15° and 30°.

Siu and Bert [20] predicted resonant frequencies and associated damping ratios for the first five modes for series of rectangular plates with free edges using the Rayleigh-Ritz method and laminated version of the Mindlin plate theory, which includes thickness-shear flexibility and rotatory and coupling inertia. Specific damping capacity in composites is predicted by Lin et al. [21] under flexural vibration using finite elements based on modal strain energy (MSE) method considering only two interlaminar stresses and neglecting transverse stress.

A variational principle to the governing equations of motion is used for the prediction of system loss factors for various modes for multilayered plates [22]. Moser and Lumassegger [23] have shown that the flexural vibrations of laminated fiber-polymer composite structures can be damped as desired by incorporation of soft ply placed optimally in zero line of longitudinal flexural stress because the additional damping is caused by shear deformation of soft ply. A high modal loss factor of 0.3–0.5 in the first mode is predicted using the finite element method.

A design methodology is presented by Sung and Snyl [24] for synthesizing a high-performance articulated manipulator fabricated with optimally tailored composite laminates. The synthesized links of the robot possess superior characteristics as high damping, high stiffness, high strength, and lower mass subject to the optimal selection of geometrical configuration, stacking sequence, laminate properties, and damping. Based on the work of Pipes and Pagano [25], Hwang and Gibson [26] developed a three-dimensional finite element/strain energy model to analyze the effect of all the three interlaminar stresses on damping in thick composite laminates under uniaxial extension. Hwang and Gibson [26] extended their previous work [27] to study the effect of vibration coupling on damping in composite beams, by decomposing the laminate energy dissipation associated with each of six resulting stress components. The contribution of the coupling-energy-dissipation terms is separately predicted to determine their effect on the total laminate-energy dissipation. The maximum contribution to damping due to the coupling effect in laminated beam occurs between 30° and 45° fiber orientations.

Saravonos and Pereira [28] developed a semi-analytical method to predict modal damping in simply supported graphite/epoxy beams. Experimentally measured and predicted the dynamic response of composite laminated plates with interlaminar damping layers further illustrated the accuracy of the method. Koo and Lee [29] used the finite element method and first-order deformation theory to study the effect of transverse shear deformation on modal loss factors and natural frequencies of composite plates. Application of complex modulus of an orthotropic lamina to modal damping and modal approach to the resultant complex eigenvalue problem-saved considerable computer time. Their observations indicate that both transverse shear and in-plane-shear increase the damping value of the composite laminate. The primary cause of delaminations or separation of the lamina in the composite laminates is the presence of transverse normal stress near the boundaries of the laminate [30]. The effect of interlaminar damping (i.e., energy dissipation by the interlaminar stresses) is supposed to be more predominant at the lamina interfaces and near the free edges especially in thick laminates.

Liao et al. [31] studied the behavior of unidirectional and symmetric angle ply carbon fiber-epoxy laminates, as well as their interleaved composites with a layer of polyethylene-co-acrylic acid (PEAA) at the midplane. Hamada [32] presented numerical and experimental investigations for the dynamic behavior of coated laminate (E-glass/polyester) composite beam. The effect of core isotropic material (steel and aluminum), natural frequency, and lamination orientation of coat on damping were studied. Coated laminated beams provide high damping in the case of aluminum core, which increases with lamina orientation of coat.

Koo and Lee [29] studied the damping of cross-ply and general laminates with off-axis laminae using the finite element method based on transverse compressibility and layer-wise distribution of in-plane displacements. It has been observed that damping analysis of thick laminates requires a refined theory taking into account the layer-wise variation of in-plane displacement because the damping is affected by the local behavior of the plate deformation.

Singh and Gupta [33] analyzed damped free vibrations of composite shells using a first-order shear deformation theory. A complex modulus approach is applied to predict system loss factors corresponding to various shell modes. Baraknov and Gassan [34] proposed a FEM/frequency dependent model based on a complex stiffness approach and laminated theory to analyze damping in laminated composite beams. Formulation of an optimized design of flexible robotic arm manipulator fabricated from fibrous polymeric composite is described by Cho [35] through the integration of design variables, such as lamina thickness, fiber orientations, and fiber volume fractions which characterize the mass, damping, and stiffness properties of a robotic manipulator. As a postdesign analysis, the sensitivity of the dynamic response of the robotic manipulator is performed to determine the most sensitive design parameter.

Singh and Kishore [36] emphasized the desirability of including damping as a constraint in their work on the damping-based design with composite materials. The case studies resulted in better dynamic performance, lower weight configurations, higher stiffness and strength of the composite structure.

Maher et al. [37] developed an improved dynamical model for vibration damping in composite structures to investigate the stacking sequence and degree of anisotropy as a function of the vibration modes. A weight factor has been introduced for correlating and updating the mathematical model to the experimental data throughout the utilization of the curve fitting response function. This has resulted in a generalized quasi-rectangular hyperbolic relationship between the loss factors and natural frequency with the confidence level at 99.5.

Yim [38] analyze the damping of composite laminates using the closed-form expression for basic damping of Poisson's ratio. Classical lamination theory based upon the elastic viscoelastic correspondence principle is utilized by developing the basic damping of Poisson's ratio for accurately predicting the damping of laminated composite beams. Basic damping of Poisson's ratio was proposed for more reliable damping prediction analysis in laminated composites by considering the relationship among the independent elastic constants along with well-known transformation equations for off-axis laminates. Damping of the symmetrically laminated composite was predicted based on the modified classical laminated theory with complex elastic constants.

Yim and Gillespie Jr. [39] determined the material damping of laminates analytically by strain energy weighted dissipation method and experimentally by the half-power bandwidth method for cantilever beam specimen excited with an impulse excitation. Unidirectional continuous fiber 0° and 90° laminates are fabricated from graphite/epoxy to examine the effect of transverse shear as a function of aspect ratio on damping of composite.

Wang and Wereley [40] presented a spectral finite element model (SFEM) for sandwich beams with passive constrained layer damping (PCLD) treatments. The viscoelastic core has a complex modulus that varies with frequency. The SFEM is formulated in the frequency domain using dynamic shape functions based on the exact displacement solutions from progressive wave methods, where they implicitly accounted for the frequency-dependent complex modulus of the viscoelastic core.

Dalenbring [41] proposed a three-dimensional material damping estimation methodology for planar isotropic material symmetry by using a constitutive viscoelastic vibration model. The proposed material model is verified, via finite-element techniques, on three laminate structures. Maheri and Adams [42] computed the modal damping of structural orthotropic materials, including anisotropic laminates, in free vibration using the Rayleigh-Ritz method. Comparison of the theoretical results has been made with experimental data obtained from tests on freely held plates, and with the results obtained from the finite element method. Kishi et al. [43] characterize the damping properties of carbon fiber-reinforced interleaved epoxy composites. Several types of thermoplastic-elastomer films, such as polyurethane elastomers, polyethylene-based isomers, and polyamide elastomers, were used as the interleaving materials. The damping properties of the composite laminates with/without the interleaf films were evaluated by the mechanical impedance method. Also, the effects of the lay-up arrangements of the carbon-fiber prepregs on the damping properties of the interleaved laminates were examined.

Nayfeh [44] developed a model for vibration parallel to the plane of lamination of the symmetric five-layer elastic–viscoelastic sandwich beam. The model is used to study the resonant frequencies and damping ratios of the lowest several modes of beams with various boundary conditions and inertia and stiffness properties as the shear stiffnesses of the viscoelastic layers are varied. Experimental results for free–free beams with contiguous and segmented constraining layers are in reasonable agreement with the predictions of the model.

Berthelot [45] analyzed the damping of orthotropic laminates with a single or two interleaved viscoelastic layers in the case of rectangular plates. The natural frequencies and modes are first derived using the Ritz method. Next, the transverse shear stresses are deduced from the fundamental equations of motion and then the transverse shear energies are obtained in each layer. Lastly, the laminate damping is evaluated from the different energies dissipated in orthotropic and viscoelastic layers. Berthelot and Sefrani [46] investigated the damping of unidirectional glass fiber composites with a single or two interleaved viscoelastic layers. The experimental damping characteristics are derived from flexural vibrations of cantilever beams as a function of the fiber orientation. The analysis considers the variations of Young's modulus and damping of viscoelastic layers with the frequency. Finally, the article presents the effects of Young's modulus and the damping of a viscoelastic layer interleaved in the middle plane of unidirectional laminate. The results are also reported for the case of external viscoelastic layers. Berthelot and Sefrani [47] developed a new model for describing the transverse damping of composites. His description of the longitudinal and transverse damping of unidirectional composites, introducing different damping coefficients associated with the motions of fibers in the longitudinal and transverse directions. The results are compared to the experimental results obtained for glass fiber, carbon fiber, and Kevlar-fiber composites.

Berthelot and Sefrani [47] carried out a damping analysis of laminate materials, laminates with interleaved viscoelastic layers, and sandwich materials. Modeling of the damping of a structure constituted of these different materials is established considering finite element analysis based on laminate theory taking into account the transverse shear effects. Experimental investigation of the damping of the different materials is implemented using beam test specimens and an impulse technique. Modeling applied to the experimental results allows us to obtain the damping parameters of the materials and constituents. Next, modeling is applied to the analysis of the damping of a simple shape structure constituted of the different materials. The results obtained are compared with the experimental results of the frequency response of the structure.

Sainsburya and Mastib [48] explore the partial coverage of cylindrical shells with a constrained viscoelastic damping layer, with emphasis on examining the minimum area of coverage that will yield optimal damping. The distribution of damping patches on the structure is based on strain energy intensity distribution maps derived for the purpose. The analysis uses the finite element method, and a suitable curved shell element is formulated for the add-on damping treatment. Numerical studies show that a partial coating procedure can be a viable approach in optimal damping designs.

Billups and Cavalli [49] compared several two-dimensional theories for predicting the specific damping capacity (SDC) of fiber composite laminates. The theories considered are those of Adams and Bacon, Adams and Maheri, Ni and Adams, and Saravanos and Chamis. No interlaminar effects are considered. Results show that the theory of Saravanos and Chamis provides the most consistent fit to experimental results for a range of material properties and laminate layups. For some laminates, however, the Saravanos and Chamis method seem to predict the wrong trend in SDC variation versus layup angle.

Prediction of damping based on the macromechanical approach in composite materials mostly relates to the effect of factors like fiber orientation, laminate configuration, frequency, and stress dependence of damping. Correspondence principle, strain energy (SE) approaches and FEM/strain energy approaches have been used for damping analysis. It has been observed that proper selection and placement of interleaving material with appropriate laminate configuration and constituent properties significantly improves the overall damping of the interleaved composite laminate.

2.3 Micromechanical approach

Micromechanical improvements in composite material damping result from changes in damping properties and geometries at or below the lamina constituent level. The micromechanical approach has been used for both continuous and short fiber-reinforced composite materials.

2.3.1 Studies on two-phase composites

Chang and Bert [50] performed a complete analytical characterization of damping and stiffness of a single layer of fiber-reinforced composite based on micromechanics. Loss tangents predicted are based upon the elastic-viscoelastic correspondence principle (for stiffness with explicit expression) and the energy approach for others. Analytical results for stiffness and loss tangents compare favorably with the experimental values for boron-epoxy, boron-aluminum, and E-glass epoxy.

White and Abdin [51] have shown that through the proper choice of fiber aspect ratio and volume fraction, damping of short aligned fiber-reinforced plastics can be improved while retaining the high stiffness. Storage and loss moduli are predicted using the Cox model [52] assuming negligible loss factor for fiber material. Experimental values of loss factor of CFRP are higher than theoretical ones due to the reason that some factors which affect its dynamic behavior are not considered in the analytical model, for example, the effect of interaction between the fibers, degree of adhesion at the interface between the fiber and matrix, the effect of shear deformation in the beam in flexure and actual stresses at the fiber ends are very high as compared to the predicted values. Further, Young's modulus in extension is not the same as in compression for CFRP composites.

Hwang and Gibson [53] proposed SE/FEM technique to predict damping and stiffness of discontinuous fiber-reinforced composites. The model is suitable to study the effect of several parameters such as aspect ratio, fiber volume fraction, fiber-matrix modulus ratio, fiber spacing, and fiber-end gap size influencing the damping properties of composite and complex structures. The resulting loss factor data from FEM is a little higher than from the modified two-phase Cox model based on a mechanics of materials analysis. In this respect, it must be noted that the FEM model is based on the actual nonuniform stress distribution whereas mechanics of materials analysis derived from the modified Cox model is based on the assumption of uniform stress throughout the composite.

Theoretical studies for damping of unidirectional CFRP (both continuous and distributed reinforcement) in longitudinal flexure, based on the law of mixture and Cox model (to consider the effect of discontinuous fibers) have been reported by Willway and White [54]. Further, damping for continuous fiber-reinforced composites in transverse flexure and transverse shear is determined using mechanics of material equations and correspondence principle with the assumption that energy dissipation occurs in resin only. The theoretical and experimental results show that it is feasible to produce unidirectional CFRP lamina having high longitudinal modulus and loss factor using a highly dissipative resin matrix.

Crema et al. [55] formulated a simple two-phase model using the energy method to predict the damping factor for kevlar fiber composite. High damping characteristics obtained for kevlar fiber composites and their independence of composite modulus indicate that kevlar fiber behaves as viscoelastic material with a damping coefficient of the same order as that of the matrix. Results show that kevlar fiber composites, especially kevlar/epoxy, has high damping and is independent of Young's modulus at least in the range of 6–70 GPa.

Much of the research on damping of composite materials has limited scope, as the studies have been restricted to individual aspects. Saravanos and Chamis [56] developed an integrated micromechanics methodology for the prediction of damping capacity for unidirectional fiber-reinforced composites. In the proposed unified approach, all six damping coefficients related to six stresses are considered. However, many other theories of micromechanical damping are restricted to damping associated with longitudinal, normal, or in-plane shear stresses. In contrast to many previous works that include only the hysteretic damping contribution of the matrix, here six damping coefficients are synthesized from hysteretic damping matrix, fiber and interface friction, and also damping due to broken fibers. Some important assumptions made are isotropic dissipative behavior for matrix and anisotropic (but transversely isotropic) for the fiber. The effect of moisture and temperature on damping in polymeric composites has also been modeled. As a result, explicit micromechanics equations based on strain energy approach relating on-axis damping capacity to fiber, matrix properties, and fiber volume fraction are obtained. Damping capacities for off-axis composites are synthesized from on-axis damping value using transformation. Transformation law to obtain off-axis damping for composite materials is proposed.

Saravanos and Chamis [57] further extended their previous work (1990) and based on it developed discrete layer damping mechanics for thick laminates, including the effects of interlaminar shear damping, and presented semi-analytical prediction of modal damping in simply supported composite plates. Application to thin laminated beams, plates, and shells demonstrates the advantage of the unified mechanics and illustrates the effect of fiber volume fraction, fiber orientation, structural geometry, and temperature on damping. Comparison of classical laminate plate damping theory with discrete layer damping theory shows that for high aspect ratio, results of both theories are comparable. But discrete layer damping theory gives better results for cross-ply laminates at low aspect ratios, higher-order modes, and high temperatures.

Das et al. [58] have suggested a model to predict damping for short fiber (distributed) reinforced bromobutyl rubber composites having a random distribution of fibers. Their approach is based on the shear lag model due to Cox [52] and Gibson et al. [59]. The study shows that the fiber efficiency is reduced by a factor of $\Pi/8$ due to the random distribution of fibers. Experimental results of damping concerning frequency and temperature are also studied.

Rikards et al. [60] developed a finite element model based on the elastic-viscoelastic correspondence principle to simulate the damping in laminated composite structures. The damping in unidirectional lamina is modeled by the unified micromechanical damping theory [61]. Considering the square packing array of fiber in the composite, damping constants $\psi_{11}, \psi_{22}, \psi_{12}$, and ψ_{23} were determined. Loss modulus and SDC in uniaxial tension, bending, and shear was predicted and compared with theoretical results [62,63].

A parallel to Saravanos and Chamis [56] work, Kaliske and Rother [64] presented a model for material damping in composite structures employing a consistent micromechanical theory of a representative fiber matrix cell due to Aboudi [65,66]. It results in six damping coefficients associated with the six engineering stress components. On a structural level, the damping is analyzed using the finite element method, that is, modeled by equivalent mass and stiffness matrices. Analogous to the mechanical properties of composites, the method results in orthotropic damping characteristics in closed form. A few basic assumptions made are: energy of composite is equal to dissipated strain energy of components, linear viscoelastic behavior of matrix and fiber, uniform state of stress in composite, no interaction stresses in material coordinates concerning damping. Further, for convenience, damping of fiber material is assumed to be isotropic as generally quoted in literature [57,66] because an experimental determination of damping values for anisotropic fiber is quite problematic. The results of Kaliske and Rother [64] model compare well with Saravanos and Chamis [56] approach.

Guan and Gibson [67] developed two analytical models for predicting damping in the woven fiber-reinforced composite. The first analytical model is closed-form solution, which involves the use of the elastic viscoelastic correspondence principle in conjunction with the elastic solution of mechanics of material theory. The second model is a three-dimensional FEM model in combination with strain energy formulation. Experiments are conducted for validation of results by the analytical model using an impulse frequency response technique to measure the stiffness and damping loss factors. The predictions from both analytical models show reasonably good agreement with the experimental results.

Chandra et al. [68] evaluated composite material damping coefficients considering damping contributions from the viscoelastic matrix, interphase, and the dissipation resulting from damage sites in various loading modes. The paper presents the results of the FEM/Strain energy investigations carried out to predict anisotropic-damping matrix comprising of loss factors η_{11}, η_{22}, η_{12}, and η_{23} considering the dissipation of energy due to fiber and matrix (two-phase) and correlate the same with various micromechanical theories.

Chandra et al. [69] presented a comparative study of a micromechanical model for the prediction of the damping coefficient of two-phase continuous fiber-reinforced composite. The effect of the shape of the fiber cross-section and fiber volume fraction on the various damping coefficient is studied through the application of the viscoelastic correspondence principle model based on Eshelby's method and Mori-Tanka approach. Adams and Maheri [70] predicted the factors that influence the stiffness and damping of laminated fiber-reinforced plastic (FRP) beams and plates using basic laminate theory and a simple damping criterion. It is also shown that the damping properties of FRP composites can be readily predicted.

Micromechanical approach to damping analysis for fiber-reinforced composites has provided better insight into the role of damping mechanisms, the effect of constituent elements (matrix, fiber), and fiber-matrix interaction in addition to the geometrical parameters. In this direction, integrated micromechanical damping theory due to Saravonos and Chamis [57] is the right step, which involves analysis and optimization for damping in fiber-reinforced composites. Chandra et al. [68,69] used FEM/strain energy approach to predict loss factors η_{22}, η_{11}, η_{12}, and η_{23} to correlate the results with other micromechanical damping models.

2.3.2 Studies on three-phase composites

Fiber-matrix interface (interfacial region) condition strongly affects the mechanical properties of composite and quite often its damping level also. Interfacial bonding in fiber-reinforced composites can be considered to be weak, ideal, or strong. The interfacial region surrounding the fibers along the length, having different properties than the properties of the main constituents phase, fiber, and matrix, is referred to as the third phase. Because of the importance of the third phase called interphase in fiber-reinforced composites, many publications related to studies on its effect on elastic and damping properties are available in the literature. Some investigations dealing with the prediction of elastic constants of composites considering the presence of interphase and their application to damping studies in three-phase fiber-reinforced composites are reviewed here.

Micromechanical methods/approaches adopted to model the composites with interphase are mostly based on mechanics of material approach, method of cells, generic unit cell method, Airy's stress function micromechanics, Eshelby's/Mori-Tanaka modified approach and FEM modeling.

A generic unit cell model based on a unique fiber substructuring concept has been proposed to predict elastic properties of composites with interphase by a group of researchers [71,72,73]. The unit cell fiber-matrix-interphase is divided into several slices and the equations are derived for each slice and integrated to obtain ply level properties. The generic model applies to both metal matrix and polymer matrix composites. Elastic moduli E_{11}, E_{22}, E_{33}, G_{12}, G_{23}, and G_{13} are predicted using this model. Kaw and Besterfield [71], De Kok et al. [72], and Chu and Rokhlin [73] modeled the interphase as springs to study its effect on the behavior of fiber-reinforced composites. Kaw and Besterfield [71] used distributed shear and normal springs to model the interphase. De Kok et al. [72] introduced springs as interphase displaying mechanical properties close to those of matrix and enabling separation of fiber and matrix since these springs represent the chemical bonds formed at the interphase as a result of chemical surface treatment. Chu and Rokhlin [73] proposed inverse determination of effective elastic moduli of fiber-matrix-interphase from experimentally measured composite moduli. Multiphase micromechanical model and simplified spring approximations are used for inversion. An analytical expression for transverse shear modulus of a composite with multi-layered fibers is derived using a multiphase generalized (MGSC) model which forms the basis of inversion.

Several publications are available which make use of the method of cells to predict the effective moduli of composites with interphase as the third constituent. Gardner et al. [74] predicted the effective properties of composites in terms of interphase material properties, interphase thickness, and fiber volume fraction. The interphase is assumed to be homogeneous with hypothetical assumptions on interphase properties being less than fiber and matrix or intermediate to fiber and matrix. Haoran et al. [75] proposed a self-consistent finite element method (SCFEM) to predict elastic moduli

(E_{11}, E_{22}, and G_{12}) of multiphase composite material with interphase for different fiber volume fractions and interphase thickness. In addition, the stress field in the interphase region is also worked out showing that E_{11} is less affected whereas E_{22} and G_{12} are affected more due to interphase thickness. Gardener et al. (1995) compared the results of the models based on the method of cells with that of the concentric cylindrical model. The model is utilized to predict interdependence amongst Young's modulus of interphase, interphase thickness, and average stress within fiber-matrix-interphase under uniaxial longitudinal tension and biaxial transverse shear. Interphase properties significantly influence the state of stress within each constituent. Low et al. [76] using the three-phase method of cells studied the effect of spatial variation of interphase on residual thermal stresses and effective properties in composites. The parameters considered are fiber volume fraction, interphase thickness, and spatial variation of interphase properties. Results suggest that the introduction of interphase gradient primarily influences the state of stress within the interphase and depending upon the interphase, stresses within may be tensile or compressive. Huang and Rokhlin [77] investigated the effect of inhomogeneous interphase on the transverse shear modulus of fiber-reinforced and particulate composites using a self-consistent (MGSC) approach and transfer matrix method.

Jayaraman et al. [78] obtained stress field in unidirectional fiber-reinforced epoxy considering a spatial variation of interphase properties using the model developed along with Mori-Tanaka analysis (nondilute local stress field). The governing equations are obtained in terms of displacements and solved in closed form. It is observed that the interphase significantly affects the local stress field, which controls the structural performance of the composite. Veazie and Qu [79] proposed an analytical estimation scheme to predict the transverse stress-strain relationship using the Mori-Tanaka method. Micromechanical stress field properties of the composite are obtained with and without partial interphase failure.

Li and Wisnom [80] applied the finite element method to analyze stress distribution in metal matrix composites with ideal interphase of uniform thickness. Achenbach and Zhu [81] investigated the effect of interphase stiffness on micro stresses of hexagonal array unidirectional fiber composite. The interphase is modeled by a layer, which resists extension and circumferential deformation. The boundary element method is used for numerical analysis of the state of stress in the basic cell keeping in view the periodicity of the basic cell.

Yi et al. [82] showed through the numerical evaluation of E_{22} that the interface significantly affects the behavior of composites. Robertson and Mall [83] performed a parametric study of fiber-matrix-interphase in high-temperature metal-matrix composites. The analytical model includes a one-dimensional mathematical model and finite element analysis using MSC/NASTRAN considering strong (perfectly bonded) and vanishing (unbonded) interfaces. For partial bonding, the model uses an elastic-plastic interphase zone between fiber and matrix.

Several review papers provide insight into theoretical and experimental work conducted on the interphase in the composites. Jayaraman et al. [84] reviewed principal interphasial characterization methods (microstructural, chemical, and mechanics) and also several recent evaluation techniques for physical characterization of (volume fraction, thickness, Young's modulus, and coefficient of thermal expansion) interphase. Jayaraman et al. [84] reviewed interphase models discussing their salient features underlying that both experimental characterization as well as modeling studies are necessary to achieve a more profound understanding of interphase, its behavior, and its effect on the properties of the composite. Scotts and Mc Cullough [85] summarized and reviewed the mechanisms of interphase formation, the role of interphasial adhesion, the influence on local stress development, and experimental studies to characterize interphase material properties. Lesko et al. [86] reviewed state of art in predicting and utilizing properties gradients of interphase in the description of composites and their micromechanics. Williams et al. [87] attempted to determine mechanical and chemical properties of the surface of single carbon fiber in an epoxy/amine matrix. The mechanism for variation of mechanical properties in interphase studies indicates that interphase is softer near the fiber due to variation of fiber bond forces as a function of fiber volume fraction.

Fisher and Brinson [88] investigate the mechanical property predictions for a three-phase viscoelastic (VE) composite by the use of two micromechanical models: the original Mori–Tanaka (MT) method and an extension of the Mori–Tanaka solution developed by Benveniste to treat fibers with interphase regions. These micro-mechanical solutions were compared to a suitable finite-element analysis, which provided the benchmark numerical results for a periodic array of inclusions. Several case studies compare the composite moduli predicted by each of these methods, highlighting the role of the interphase.

A micromechanical model of Young's modulus for three special cases of composites—continuous unidirectional fiber composites, particulate composites, and periodically bi-laminate composites—has been developed by Lim et al. [89] within the framework of the mechanic-of-materials approach. Parallel-series (PS) and series-parallel (SP) schemes were adopted based upon the opposing combination sequence of the parallel model (direct rule-of-mixture) and the series model (inverse rule-of-mixture). By expressing the inclusion's geometrical characteristics in terms of reinforcement volume fraction and interphase volume, a set of reinforcement parameters for the inclusion and its interphase are extracted. Based on reinforcement parameters for the special cases, those of more general cases, which are intermediates of the special cases, are obtained by curve-fitting approximation. Young's modulus of simple cases such as coated fiber composites in the longitudinal direction and coated lamina composites are shown to be approximate to the parallel and series models respectively.

Xun et al. [90] analyzed the overall elastic and viscoelastic properties of particulate composites with graded interphase analytically. The localization relation for a coated particle with graded interphase in the radial direction has been firstly derived, and the effective elastic moduli of such composites are then estimated by the mean-field theory and generalized self-consistent method respectively. The effective storage and loss moduli of the composite are also determined through the dynamic correspondence principle. The results show that the nature of the interphase (soft and hard) can have an important influence on the prediction for effective elastic and viscoelastic properties of the composite, especially for a soft graded interphase, the predictions based on a uniform interphase model overestimates largely the effective elastic and viscoelastic moduli of the composite.

Wang et al. [91] describe a numerical approach for modeling the micromechanical behavior and macroscopic properties of multiphase fiber-reinforced composites with inhomogeneous interphases. The interphases are modeled as functionally graded elastic layers with the Youngs modulus and Poissons ratio varying in the radial direction. In general, the fibers can have different elastic properties and sizes and can, if desired, be randomly distributed. The approach is based on the numerical solution of a complex boundary integral equation in which the boundary parameters are expressed in terms of complex Fourier series. All the integration has been done analytically and thus the method allows for accurate calculation of the elastic fields anywhere within the material, including inside the fibers and interphase. Explicit expressions for the effective elastic constants can be obtained from general relations between the average stresses and strains.

Some investigations on modeling the effect of interphase on damping in fiber-reinforced composites are discussed further. The ideal interface plays the role of transferring load and does not contribute to damping. Nielson [92] introduced the rule of mixtures to determine loss tangent (tan $\varphi^{\circ}{}_{c}$) for ideal composites. Effect of type of interfacial bonding covering the range from weak to strong is considered in the works of Kubat [93] and Ziegel and Ramanov [94]. They proposed two parameters A and B, covering together a whole range of interfacial imperfections. Parameter A (relative damping at the interface to that of the matrix), and interfacial damping (tan φ_{i}) characterize the interfacial bonding in composite with poor adhesion, whereas strong interfacial region (that do not contribute to damping) is defined by parameter B. Murayama [95] developed a model to predict damping (as a difference between the damping of real and ideal composite) at the interface of nylon and PET fiber-reinforced vulcanized rubber composites and observed direct relation to interfacial shear strength. Similarly, Chinquin et al. [96] observed a decrease in damping associated with improvement in interface bonding. These observations have been validated by the dynamic mechanical analysis (DMA) tests, which provide greater accuracy for the measurement of interfacial damping than any other mechanical tests.

Chaturvedi and Tzeng [97] developed three micromechanical models for the prediction of damping in aligned short fiber-reinforced three-phase polymer composites. These models are more comprehensive and exact as compared to simple models referred to above [92,98]. Here, the interphase is referred to as a distinct third phase between the fiber and bulk matrix with its viscoelastic properties. Parametric studies indicate that the fiber aspect ratio (l/d), the elastic modulus and damping properties of the interphase material appear to be dominant parameters that significantly influence the dynamic stiffness and viscoelastic damping properties of the three-phase composite system. The main drawback of the model is that the input data for the interphase, that is, geometric and material properties of interphase have been interpolated [99] and no experimental values are available. In some cases, the volume fraction of interphase has been determined based on the hexagonal array of fiber surrounded by a modified matrix layer resulting in an interphase volume fraction (V_I) varying between 5 and 30%. Vantomme [100] developed the two-phase and three-phase models using an energy balance approach to obtain a closed form relationship between material properties and design parameters. Longitudinal, transverse, and shear loss factors for the composite material (E-glass fiber epoxy) are predicted for each mode in terms of the percentage contribution of damping due to fiber, matrix, and interphase. The three-phase model shows that a poor quality interface with low elastic stiffness has a significant effect on the damping capacity of unidirectional laminate. However, for a normal quality interface, more sensitive and accurate damping measurement methods are needed to identify the contribution of interface damping to the overall damping. Finegan and Gibson [101] proposed a strain energy/FEM model to study the contribution of interphase to overall damping in coated fiber-reinforced composites.

Finegan and Gibson [102] develop an analytical model of damping at the micromechanical level in polymer composite reinforcement will coated fibers using the theory of elasticity, finite element, and mechanics of material. Micromechanics equations are presented to define the damping and stiffness of unidirectional coated fiber composites under longitudinal normal (LN), transverse normal (TN), transverse shear (TS), and longitudinal shear (LS) loading conditions. It is shown that the use of viscoelastic polymer coating on the fibers is an effective way of improving damping in composites.

Chandra et al. [68] calculated damping in a three-phase (i.e., fiber-interphase-matrix) composite as an attempt to understand the effect of interphase. Parametric studies on damping for the three-phase composite are carried out and, which have shown that the change in properties of fiber, matrix, and interphase leads to a change in the magnitude of the effectiveness of interphase, but how interphase would affect the various loss factors depends predominately upon whether the hard or soft interphase is chosen.

Gu et al. [103] proposed a new model for calculating the damping capacity of particulate-reinforced metal matrix composites (PMMCs) is proposed from the viewpoint of energy dissipation. The finite element method (FEM) has been employed to investigate the effect of interphase on the damping capacity of composite by varying the thickness and material properties of the interphase region. The calculated results show that the damping of composites containing the elastic interphase increases with the elastic modulus. For the case of plastic interphase, the weak interphase or the hard interphase is beneficial to improve the overall damping of composites.

Concerted efforts are needed to develop three-phase micromechanical damping models to incorporate interphase effect, fiber packing under various loading conditions. Experimental characterization of interphase elastic and damping properties is another major problem, which has not been investigated in detail. A comparative study of damping in three-phase composites using various models and parametric study as regard to the effect of volume fraction of interphase, nature of interphase, and frequency is also desirable.

2.3.3 Viscoelastic approach

The viscoelastic approach has been used successfully by several researchers to study the dynamic behavior of viscoelastic materials. Using damping measurements, many internal variables and relaxation mechanisms are identified that range in geometrical scale from crystal lattice dimensions to structural dimensions and in temporal scale over a similarly broad range [104]. Bagley and Torvik [105] modeled the dynamic behavior of viscoelastically damped structures by fractional calculus approach over a wide range of frequency providing a closed-form solution to equations of motion.

A finite element formulation to incorporate the effect of damping of viscoelastic material is studied by Golla and Hughes [106] through the use of dissipation coordinates to canonical M and K form of undamped motion equations. Lesieture and Mingori [107] developed augmenting thermodynamic field (ATF) model, a time-domain continuum model of material damping that preserves characteristic frequency-dependent behavior and compatibility with finite element structural analysis methods.

Lesieture and Mingori [108] developed an anelastic displacement field (ADF) approach to model the time-domain longitudinal behavior of unidirectional long fiber composite made from linear viscoelastic constituents. Here, instead of addressing physical damping mechanisms directly as in the earlier approach, their effect on the displacement field is considered. In this approach, the total displacement is considered to comprise of two parts: an elastic part and an anelastic part. Effective composite anelastic displacement fields and material properties (frequency and time-dependent composite modulus and composite damping) are obtained in terms of corresponding constituent properties and respective volume fractions.

Pritz [109] proposed a four-parameter fractional derivative model applicable to describe damping properties of materials with low vibration damping (metals) as well as high damping polymers over a wide frequency range.

Li and Weng [110] reported a general micromechanical viscoelastic theory using Burger's four-parameter model to study the effective creep behavior and nine complex parameters of a viscoelastic orthotropic composite consisting of a viscoelastic matrix and aligned elliptical fibers. The real and imaginary parts of nine complex moduli under harmonic loading at low frequency are examined as a function of aspect ratio, the volume fraction of fibers, and frequency. The real part of effective moduli correlated well with their elastic counterparts with an increase in frequency, whereas the imaginary parts are found to be dependent upon the cross-sectional shape and also the loading frequency.

Tao et al. [111] presented a new method for calculating the viscoelastic damping matrix of fiber composite. Firstly for lamina energy dissipation, the formula is obtained by using damping matrix and theory of linear anisotropic viscoelasticity. Three-D effective damping loss for laminate is obtained by using the energy method. Based on this model the energy loss of arbitrary composite under complex loading situations can be calculated. This method has the advantages of generality, simplicity, and directness.

Rao [112] describes the application of passive damping technology using viscoelastic materials to control noise and vibration in vehicles and commercial airplanes. Special damped laminates and spray paints suitable for mass production and capable of forming with conventional techniques are now manufactured in a continuous manner using advanced processes. These are widely used in the automotive and aerospace industry in a variety of applications to reduce noise and vibration and to improve interior sound quality. Patel et al. [113] devise a simple finite element based unit cell model to calculate the effective tanδ of a composite as a function of frequency. Using this method, they show that if the relaxation times of the constituents are properly chosen, a flat tanδ response over a wide frequency range can be obtained. Inclusions with multiple layers are seen to be particularly suitable towards this end.

2.4 Nonlinear damping

Nonlinear behavior in structural dynamics is somewhat difficult to define in a general manner. From an engineering standpoint, a system is considered to be nonlinear if its dynamic properties, such as stiffness and/or damping depend on displacement, velocity, acceleration, or any combination of these variables. The general problem of studying the dynamics of a nonlinear system is further compounded by the fact that nonlinear behavior cannot be isolated from the operating range and conditions that give rise to it in the first place. Considering that such conditions may encompass a wide range of situations, a complete analysis, theoretical or experimental, may become very expensive, if not impracticable. Consequently, nonlinear effects are usually ignored even though most practical engineering structures behave in a nonlinear fashion, under certain operating conditions. The detection, identification, and quantification of structural nonlinearities have been the subject of many research papers. In this wider industrial world, the composite is replacing more traditional materials but design teams require information on the behavior of the material under the conditions that it will experience in the applications envisaged. Often the structures subjected to high loads and much greater cyclic strains are required to model.

Many researchers have modeled the linear and nonlinear damping like Adams and Maheri [114] predicted moduli and flexural damping of anisotropic CFRP and GFP beams concerning fiber orientation using basic elastic relationships for unidirectional composites together with the Adams-Bacon criterion. The effect of aspect ratio and stress level on damping is also considered. Results on the effect of stress amplitude on specific damping capacity (SDC) for unidirectional CFRP and GFRP with different fiber angle beams are reported. For the low angle of beams (0° and 15°), SDC is seen to be independent over the whole stress range, that is, the composite behaves as a Hookean elastic material and the energy loop is elliptical, the energy is proportional to the square of the amplitude. For higher angle beams (30–90°), SDC is governed by fiber-resin interaction, that is, the onset of nonlinearity of damping in CFRP begins at about 1 MPa, and while in GFRP there is reduced nonlinearity for stress up to 4 MPa. The nonlinearity of SDC of material can be attributed to plastic deformation beyond certain critical stress levels. It has been observed that in polymeric reinforced composites, nonlinear behavior is more significant in shear and transverse direction to fiber due to considerable straining of resin. In addition, material and structural integrity determined by defects such as microstructure (voids, impurities, and imperfections in resin/fiber bonding or debonding) are the points of stress concentration that increase the damping and add to nonlinearity.

Kenny and Marchetti [115] proposed a mathematical model, which could predict the viscoelastic-plastic response of the material at high stresses and its influence on fatigue damage in PEEK (poly ether ketone) matrix carbon fiber composites. Good agreements between theoretical and experimental results are observed indicating discontinuity in damping at yield stress and high values of specific damping at high stresses (plastic damping). On the other hand, a very high rise in temperature is noticed in regions of plastic damping. Orth et al. [116] proposed a new method for measuring hysteretic damping during dynamic testing which allows four different properties: stress, extension, stiffness, and mechanical energies to be measured simultaneously. For this, a test program controlled by a computer utilizing a function generator card to control the test machine and data acquisition and calculation of hysteretic properties has been developed. A wind turbine blade (GFRE) and spring retainer (CFRP) so tested indicates sensitiveness of damping to damage. Mantena et al. [117] employed damping measurements obtained from the hysteresis loop technique to justify its usefulness as an inspection tool for evaluation of damage due to manufacturing parameters in the pultruded products (beams, trusses, stiffeners, etc.). Mantena et al. [118] used the vibration technique (dynamic mechanical analyzer) to verify the state of pultruded samples with induced contaminations (simulated porosity and interfacial debonding) and compared the results with ultrasonic measurements. The sensitivity of the response of the two methods to fiber fraction changes and manufacturing parameters shows the potential of these methods for online nondestructive quality control/monitoring (both global and local methods) of pultruded products made from composite materials. The loss factor of vibration testing compared with ultrasonic loss factor from ultrasonic tests, both being a measure of material damping are used to quantify the damage. For this purpose, the percentage change in peak loss factor of each contaminated sample from the uncontaminated sample is computed. Based on the response and the manufacturing variables, a statistical regression analysis procedure is evolved to quantify damage in composites.

Nelson and Hancock [119] developed a simple model to determine the amount of energy dissipated by the fiber-matrix interface due to sliding and the viscoelastic matrix. Interfacial slip shows a marked effect on the stress-strain behavior of rubbery polymer reinforced with short fibers. Saravanos and Hopkins [120] based on generalized laminate theory developed a unified inclusive laminate theory for stiffness, damping, and inertia terms, which can either handle pristine or delaminated composite laminates with single or multiple delaminations. The kinematic assumptions allow for in-plane and out-of-plane relative motion between the delaminated surfaces and apply to general lamination configuration. Analytical results of natural frequencies, modes, and modal damping in graphite-epoxy composite beams with central delamination are correlated with the experimental measurements. There is a very good agreement for the case of natural frequencies and a fair agreement for the modal damping. The effects of delamination on damping are more dependent on the laminate configuration. Rapid and significant damping changes may provide a reliable indication of delamination damage and growth in connection with a decrease in natural frequencies. It is underlined that natural frequencies are rather insensitive to interfacial friction and interfacial damping appears to be predominant at large delaminations.

Kinra et al. [121] measure the nonlinear intrinsic material damping in continuous fiber-reinforced metal matrix composite. The particular MMC studied is a four-ply Al composite with $\theta = 0, 15, 30, 45, 60, 75, 90°$. Free decay of flexural vibrations of beam method is used for damping measurement. Specific damping has been obtained at a different angle of fiber for varying strain amplitude. Chandra et al. [68] developed a pseudo-dynamic model based on the FEM/strain energy approach to study the contribution of energy dissipation due to sliding at the fiber-matrix interface to evaluate its effect on η_{11}, η_{22}, η_{12}, and η_{23} in a fiber-reinforced composite having damage in the form of hairline debonding. Analysis of the effect of damage on composite damping indicates that it is sensitive to its orientation and type of loading. Zhang and Hartwig [122] uses the damping measurement for the investigation of damage processes by fatigue cycles of unidirectional (UD) fiber composites. UD AS4 carbon fiber reinforced PEEK and epoxy, as well as fiberglass reinforced epoxy, were considered. A "damping plateau" has been primarily detected from the damping dependence on fatigue cycles in epoxy composites, and the mechanism of the damping plateau is discussed under the assumption of possible energy balance between fatigue load input and damage dissipation. The fatigue life of a UD composite with a brittle epoxide matrix is better than that with a ductile thermoplastic matrix by using the same fiber reinforcement. The formation and growth of microcracks in the epoxide matrix contributes to the energy dissipation under fatigue load, which seemly improves the fatigue behavior. Damping is recommended in the evaluation of the damage process, which seems more sensitive than stiffness.

Kyriazoglou and Guild [123] develop a hybrid methodology for evaluating the effect of damage on the damping properties of woven damaged woven and nonwoven Glass Fiber Reinforced Plastics (GFRP) and Carbon Fiber Reinforced Plastics (CFRP) laminates. The main damage mechanisms considered are matrix cracking and tow fracture. Furthermore, the effect of highly localized damage mechanisms on damping is analyzed, in the case of woven GFRP and CFRP laminates containing a circular notch. This hybrid methodology is a synergy of laboratory observations and Finite Element Analysis (FEA) that eventually leads to the evaluation of the effect of homogeneously distributed damage to a composite laminates damped response, analysis of the vibration of notched laminates that cannot be provided for by continuum mechanics and the analysis of localized damage mechanisms in woven CFRP laminates and generally systems that do not facilitate visual inspection of damage mechanisms.

Yi et al. [124] investigated thermally induced stresses that occur during the cool-down process after curing based on nonlinear thermo-viscoelastic finite element methods. Nonlinear viscoelastic stress singularities near free edges of laminated composites are evaluated as a function of time and cool-down history. Various cool-down histories are considered. Temperature distributions inside composites are predicted by solving heat conduction equations. Numerical results indicate a strong sensitivity to the nonlinearities of viscoelastic constitutive relations. Temperature history significantly affects residual interlaminar and in-plane stresses. The rate of stress relaxation also varies with the loading history. Sokolinsky et al. [125] demonstrated the practical value of the geometrically nonlinear higher-order theory is using four-point bend tests carried out on sandwich beam specimens comprised of aluminum face sheets and a PVC foam core. The experimental results were compared with the predictions of the classical sandwich theory, and with linear and geometrically nonlinear higher-order sandwich panel theory. The analytical predictions based on the higher-order theory are in excellent agreement with the experimental results. Response parameters show fundamentally distinct behavior with increasing external load, both in the particular section and along the span. Considering the longitudinal displacements, there is a significant geometrically nonlinear stage of the response that precedes the appearance of the material nonlinearity. The peeling stresses also exhibit significant geometrical nonlinearity in the vicinity of the internal supports. The linear higher-order theory can be used efficiently to estimate the vertical displacements of the soft-core sandwich beams up to high load levels with great accuracy. Spathis et al. [126] studied the nonlinear behavior of polymeric fiber composite materials theoretically in terms of a model developed for elastic-plastic materials, and generally valid for an elastic-plastic response. This theory is focused on the current material state and connects material anisotropic response with identifiable directions in the present material state. The tensile stress-strain response of unidirectional fiber-reinforced polymer composite, at various angles in respect of fiber angle, has been obtained. The strain rate effect has been depicted using a scaling law, obtained in viscoelasticity and introduced in the functional form of rate of plastic deformation. Pooler and Smith [127] attempted to describe the time and temperature dependence of a commercial WPC formulation. Characterization of the formulation was undertaken using a series of creep and recovery tests. A Prony Series was used to describe the material's time-dependent compliance, where time was shifted with stress and temperature to describe the observed nonlinear response and temperature dependence. Damage effects were successfully correlated isothermally using effective stress.

Paepegem et al. [128] explains the development of a material model to predict mechanical behavior and discusses the experimental program with tensile tests on $[+45/45]_{2s}$ laminates and off-axis $[10]_{8s}$ composites. Cyclic tensile tests have been performed to assess the amount of permanent shear strain and the residual shear modulus. Pramanick and Sain [129] develop a generic creep prediction model that describes the creep behavior of composites with the constituents' creep behavior. The 'theory of mixture' for composites is extended to describe the creep behavior of this material, which is two-phase. This model is validated for HDPE-rice husk composites with power-law-Boltzmann's superposition principles. The model works well not only to describe creep, but also its nonlinearity. The model is generic enough for extending it to incorporate varying environmental conditions, such as time and temperature. Ellyin and Xia [130] presented a nonlinear viscoelastic constitutive model, in differential form, based on the deformation characteristics of thermoset polymers under complex loadings. This rheological model includes a criterion to delineate loading and unloading in multiaxial stress states and different moduli for loading and unloading behaviors. The material constants and functions of this model are calibrated following a well-defined procedure. The model predictions are compared with the experimental data of an epoxy polymer subjected to uniaxial and biaxial stress states with monotonic and cyclic loading. The agreement is very good for various loading regimes. The constitutive model is further implemented in a finite element code and the residual stresses arising from the curing process of polymer reinforced composites are determined for two different epoxy resins.

Micromechanical models for a study of nonlinear viscoelastic (NVE) response of composite laminae are developed by Sridharan [131] and their performance is compared. A single integral constitutive law proposed by Schapery and subsequently generalized to multiaxial states of stress is utilized in the study for the matrix material. Elastic moduli for composite have been predicted and variation of strain with time has been obtained. Han et al. [132] presented the formulation of a nonlinear composite 9-node modified first-order shear deformable element-based Lagrangian shell element for the solution of geometrically nonlinear analysis of laminated composite thin plates and shells. The application limit of modified shear deformation theory is presented for the correct analysis of composite laminates. However, it is evident that it results in a parabolic distribution of the transverse shear strains and satisfies the zero transverse shear stresses requirements at the shell surfaces. It also requires insignificant modifications to be implemented in existing displacement-based first-order shell elements. Natural co-ordinate-based higher-order transverse shear strains are used in the present shell element.

Most of the work on nonlinear damping is related to the stress and strain evaluation considering the effect of temperature and strain amplitude. Cyclic loading condition is not discussed in detail. There is a need for the evaluation of damping for cyclic loading and the effect of frequency on loss factor.

2.5 Conclusion

Good dynamic properties of the fiber-reinforced composite can be tailored by proper selection of its constituent property as discussed in the literature. The following conclusions are drawn based on the comprehensive literature review.

a) Analytical prediction of damping in linear viscoelastic fiber-reinforced composite materials/laminates is mostly carried out by application of viscoelastic correspondence principle and strain energy approach.

b) Most of the work available considers only square packing for the prediction of damping. There is a need to correlate the fiber packing factor with the loss factor.

c) Damping evaluation of three-phase composite is not widely covered although some work is available on evaluation of material damping along fiber considering the effect of fiber-matrix-interphase. Development of a three-phase model and comparative study with existing models and also the parametric study is desirable.

d) Some work on the application of the micromechanical damping approach to obtain high damping through interleaving technique and hybrid composites is reported in the literature. However, an integrated approach comprising of techniques of improvement of damping and damping tailoring, good mechanical properties, and low weight can provide composite materials/structures with better dynamic performance.

e) Optimization for damping can improve the dynamic performance of laminated composites since damping is anisotropic, highly tailorable, and depends on an array of micromechanical, laminate, structural parameters.

Thus optimization for damping in fiber-reinforced composite structures requires further studies on multiple design criteria approach for different applications.

f) Most of the work on nonlinear damping is based on a micromechanical approach hence there is a need to model the nonlinearity on the micromechanical approach. Very limited works are available considering cyclic loading for the prediction of the dynamic performance of the fiber-reinforced composite.

g) Limited works are available which consider the effect of amplitude of vibration on damping. There is a need to study the effect of amplitude of vibration on damping in composites both in the linear and nonlinear range, as these composites have good dynamic properties.

References

[1] P. Zioniev, Y.N. Ermkov, Energy Dissipation in Composite Materials, Tchnomic Publication Co.Inc., 1994.
[2] D.J. Nelson, J.W. Hancock, Interfacial Slip and Damping in Fiber Reinforced Composite, Journal of Mater. Science 13 (1978) 2429–2440.
[3] R. Chandra, in: A study of Behaviour of Fibre-Rienforced Composites Proceedings of workshop on Solid Mechanics, India, University of Roorkee, 1985, pp. 59–63.
[4] J.M. Kenny, M. Marchetti, Elasto Plastic Behavior of Thermoplastic Composite Laminates under Cyclic Loading, Composite Structures 32 (1995) 375–382.
[5] J.M. Curtis, D.R. Moore, Fatigue Testing of Multiangle Laminates of CF/PEEK, Composites 19 (1988) 446–455.
[6] Z. Hashin, Complex Moduli of Viscoelastic Composites: I General Theory and Application to Particulate Composites, International Journal of Solids and Structures 6 (1970) 539–552.
[7] S.A. Suarez, R.F. Gibson, C.T. Sun, S.K. Chaturvedi, The Influence of Fiber Length and Fiber Orientation on Damping and Stiffness of Polymer Composite Materials, Experimental Mechanics 26 (2) (1986) 175–184.
[8] C.T. Sun, J.K. Wu, R.F. Gibson, Prediction of Material Damping in Randomly Oriented Short Fiber Polymer Matrix Composites, Journal of Reinforced Plastics and Composites 4 (1985) 262–272.
[9] C.T. Sun, J.K. Wu, R.F. Gibson, Prediction of Material Damping of Laminated Polymer Matrix Composites, Journal of Material Science 22 (1987) 1006–1012.
[10] C.W. Laird, H.B. Kingsbury, A Method of Determining Complex Moduli of Viscoelastic Materials, Experimental Mechanics 13 (2) (1973) 126.
[11] E.B. Paxon Jr., Real and Imaginary Parts of Complex Viscoelastic Modulus for Boron-Fiber-Reinforced Plastics, Journal of Acoustical Society of America 57 (4) (1975) 891.
[12] R.F. Gibson, Dynamic Mechanical Behavior of Fiber – Reinforced Composite: Measurements and Analysis, Journal Composite Materials 10 (1976) 325.
[13] E. Rosenblueth, I. Herrera, On a kind of hysteretic damping, Journal of Engineering Mechanics Division ASCE. 90. (1964) 37–48, doi:10.1061/JMCEA3.0000643.
[14] R.D. Adams, M.A.O. Fox, R.J.L. Flood, R.J. Friend, R.L. Hewitt, The dynamic properties of unidirectional carbon and glass fiber reinforced plastics in torsion and flexure, Journal of Composite Material 3 (1969) 594–603 .2.

[15] R.D. Adams, D.G.C. Bacon, Measurement of Flexural Dynamic Capacity and Dynamic Young's Modulus of metals and Reinforced Plastics, Journal of Physics D: Applied Physics 6 (1973) 27–41.
[16] R.G. Ni, D.N. Lin, R.D. Adams, The Dynamic Properties of Carbon-Glass Fiber Sandwich Laminated Composites: Theoretical Experimental and Economic Considerations, Composites 15 (4) (1984) 297–304.
[17] D.X. Lin, R.G. Ni, R.D. Adams, Prediction and Measurement of The Vibrational Damping Parameter of Carbon Glass Fiber Reinforced Plastic Plates, Journal of Composite Materials 18 (1984) 132–152.
[18] M.E. Mcintyre, J. Woodhouse, On Measuring the Elastic and Damping Constants of Orthotropic Sheet Materials, Acta Metalurgica 36 (6) (1988) 1397–1416.
[19] R.M. Crane, J.W. Gillespie, Characterization of the Vibration Damping Loss Factor of Glass and Graphite Fiber Composites, Comp Science Technology 40 (1991) 355–375.
[20] C.C. Siu, C.W. Bert, Sinusoidal Response of Composite Material Plates With Material Damping, Transaction of ASME, Journal of Engineering for Industry 96 (1974) 603–610.
[21] D.X. Lin, R.G. Ni, R.D. Adams, Prediction and Measurement of The Vibrational Damping Parameter of Carbon Glass Fiber Reinforced Plastic Plates, Journal of Composite Materials 18 (1984) 132–152.
[22] N. Alam, N.T. Asnani, Vibration and Damping Analysis of Fiber-Reinforced Composite Plates, Journal of Composite Materials 20 (1986) 2–18.
[23] K. Moser, M. Lumasseggar, Increasing the Damping of Flexural Vibration of Laminated FPC Structures by Incorporation of Soft Intermediate Plies With Minimum Reduction of Stiffness, Composite Structures 10 (1988) 321–333.
[24] C.K. Sung, S.S. Snyl, Composite Laminated Box-Section Beam Design for Obtaining Optimal Elasto-Dynamic Responses of Flexible Robot Manipulator, International Journal of Mechanical Science 32 (5) (1990) 391–404.
[25] R.B. Pipes, N.J. Pagano, Interlaminar Stresses in Composite Laminates under Uniform Extension, Journal of Composite Materials 4 (1970) 538–548.
[26] S.J. Hwang, R.F. Gibson, Contribution of Interlaminar Stresses to Damping in Thick Laminated Composites under Uniaxial Extension, Composite Structures 20 (1992) 29–35.
[27] S.J. Hwang, R.F. Gibson, Micromechanical Modeling of Damping in Discontinuous Fiber Composite using a strain Energy /Finite Element Approach, Journal of Material Science and Technology 109 (1987) 47–52.
[28] D.A. Saravonos, J.M. Pereira, Effects of Interply Damping Layers on the Dynamic Characteristics of Composite Plates, AIAA Journal 30 (12) (1992) 2906–2913.
[29] K.N. Koo, I.A. Lee, Refined Analysis of Vibration and Damping for Anisotropic Laminates in Cylindrical Bending, Journal of Sound and Vibration 184 (4) (1995) 553–566.
[30] R.F. Gibson, Principles of Composite Mechanics, McGraw-Hill Inc., 1994.
[31] F.S. Liao, A.C. Su, T.C. Hsu, Vibration Damping of Interleaved Carbon Fiber-Epoxy Composite Beams, Journal of Composite Materials 28 (18) (1994) 1840–1854.
[32] A. Hamada, A.E. Hamid, Vibration and Damping Analysis of Beams with Composite Coats, Composite Structures 32 (1995) 33–38.
[33] S.P. Singh, K. Gupta, Damped Free Vibrations of Layered Composite Cylindrical Shells, Journal of Sound and Vibration 172 (2) (1994) 191–209.
[34] R.G. Baraknov, J. Gassan, Frequency Response Analysis of Laminated Composite Beams, Mechanics of Composite Materials 30 (5) (1994) 664–674.
[35] L.P. Cho, Optimal Design and Sensitivity Analysis of Flexible Robotics Manipulator Fabricated from Advanced Composite Materials, Journal of Thermoplastic Composite Materials 8 (4) (1995) 346–364.

[36] S.P. Singh, A.K. Kishore, Damping Based Design With Composite Materials, in: Proc. IUTAM-IIT Delhi, Int. Winter School on Optimum Dynamic Design Using Modal Testing and Structural Dynamic Modification, 15-19 Dec., India, 1997, pp. 445–453.
[37] A. Maher, F. Ramadan, M. Ferra, Modeling Of Vibration Damping In Composite Structures, Composites Structures 46 (1999) 163–170.
[38] J.H. Yim, A Damping Analysis of Composite Laminates Using the Closed Form Expression for Basic Damping of Poisson's Ratio, Composite Structures 46 (1999) 405–411.
[39] J.H. Yim, J.W. Gillespie, Damping Characteristics of 0 and 90 AS4/3501-6 Unidirectional Laminates Including the Transverse Shear Effect, Composite Structures 50 (2000) 225–227.
[40] G. Wang, N.M. Wereley, Spectral Finite Element Analysis of Sand Witch Beams with Passive Constrained Layer Damping, Journal of Vibration and Acoustics 124 (2002) 376–386.
[41] M. Dalenbring, Experimental Material Damping Estimation for Planar Isotropic Laminate Structures, International Journal of Solids and Structures 39 (2002) 5053–5079.
[42] M.R. Maheri, R.D. Adams, Modal Vibration Damping of Anisotropic FRP Laminates using The Rayleigh–Ritz Energy Minimization Scheme, Journal of Sound and Vibration 259 (1) (2003) 17–29.
[43] H. Kishi, M. Kuwata, S. Matsuda, T. Asami, A. Murakami, Damping Properties of Thermoplastic-Elastomer Interleaved Carbon Fiber-Reinforced Epoxy Composites, Composites Science and Technology 64 (16) (2004) 2517–2523.
[44] S.A. Nayfeh, Damping of Flexural Vibration in the Plane of Lamination of Elastic–Viscoelastic Sandwich Beams, Journal of Sound and Vibration 276 (2004) 689–711.
[45] J.M Berthelot, Damping Analysis of Orthotropic Composites with Interleaved Viscoelastic Layers: Modeling, Journal of Composite Materials 40 (2006) 1889–1909.
[46] J.M. Berthelot, Y. Sefrani, Damping Analysis of Unidirectional Glass Fiber Composites with Interleaved Viscoelastic Layers: Experimental Investigation and Discussion, Journal of Composite Materials 40 (2006) 1911–1932.
[47] J.M. Berthelot, Y. Sefrani, Longitudinal and Transverse Damping of Unidirectional Fibre Composites, Composite Structures 79 (2007) 423–431.
[48] M.G. Sainsburya, R.S. Mastib, Vibration Damping of Cylindrical Shells using Strain-Energy-Based Distribution of an Add-on Viscoelastic Treatment, Finite Elements in Analysis and Design 43 (2007) 175–192.
[49] E.K. Billups, M.N Cavalli, 2D Damping Predictions of Fiber Composite Plates: Layup Effects, Composites Science and Technology 68 (2008) 727–733.
[50] S Chang, C.W Bert, Analysis of Damping for Filamentary Composite material, in: Proceedings of Sixth St.Louis Symposium, American Society of Metals, 11-12 May 1973, pp. 51–62.
[51] R.G. White, E.M.Y. Abdin, Dynamic Properties of Aligned Short Carbon Fiber Reinforced Plastics in Flexure and Torsion, Composite 16 (4) (1985) 293–306.
[52] H.L. Cox, The Elasticity and Strength of paper and other Fibrous Materials, British Journal of Applied Physics 3 (1952) 72–79.
[53] S.J. Hwang, R.F. Gibson, Micromechanical Modeling of Damping in Discontinuous Fiber Composite using a strain Energy /Finite Element Approach, Journal of Material Science and Technology 109 (1987) 47–52.
[54] T.A. Willway, R.G. White, The Effect of Matrix Complex Modulus on Damping Properties of CFRP Laminae, Composite Science and Technology 36 (1989) 77–94.
[55] L.B. Crema, A. Castellani, U. Drago, Damping Characteristics of Fabric and Laminated Kevlar Composite, Composites 20 (6) (1989) 593–596.

[56] D.A. Saravanos, C.C. Chamis, Unified Micromechanics of Damping for Unidirectional and Off-Axis Fiber Composites, Journal of Composite Technology and Research 12 (1) (1990) 31–40.
[57] D.A. Saravanos, C.C. Chamis, Integrated Mechanics For Passive Damping of Polymer Matrix Composites and Composite Structures, in: Mechanics and Mechanisms of Material Damping Conference Sponsored By ASTM, Baltimore, Maryland, 13-15 March 1991, pp. 1–25.
[58] J.N. Das, N.G. Nair, N. Subramaniam, In-Plane and Transverse Damping Characteristics of Short-Fiber Reinforced Bromobutyl Rubber Composites, Rubber and Composites: Processing and Applications 22 (1993) 249–255.
[59] R.F. Gibson, S.K. Charturvedi, C.T. Sun, Complex Moduli of Aligned Discontinuous Fiber-Reinforced Polymer Composites, Journal of Material Science 17 (1982) 3499–3509.
[60] R. Rikards, A. Chate, A.K. Bliedzki, V. Kushnevsky, Numerical Modeling of Damping Properties of Laminated Composite, Mechanics of Composite Materials 30 (3) (1994) 359–371.
[61] D.A. Saravanos, C.C. Chamis, Unified Micromechanics of Damping for Unidirectional and Off-Axis Fiber Composites, Journal of Composite Technology and Research 12 (1) (1990) 31–40.
[62] R.M. Crane, Jr Gillespe, Analytical Model for Prediction of the Damping Loss Factor of Composite Materials, Polymer Composite 13 (3) (1992) 179–190.
[63] S Chang, C.W Bert, Analysis of Damping for Filamentary Composite material, in: Proceedings of Sixth St.Louis Symposium, American Society of Metals, 11-12 May 1973, pp. 51–62.
[64] M. Kaliske, H. Rother, Damping Characterization of Unidirectional Fiber Reinforced Composite, Composite Engineering 5 (5) (1995) 551–567.
[65] J. Abuodi, A Continuum Theory for Fiber-Reinforced Elastic Viscoelastic Composites, International Journal of Engineering Science 20 (1982) 605–621.
[66] J. Aboudi, Micromechanica Analysis of Composites by Method of Cells, Applied Mechanics Reviews 42 (1989) 193–221.
[67] H. Guan, R.F. Gibson, Micromechanical Models for Damping in Woven Fabric-Reinforced Polymer Matrix Composites, Journal of Composite Materials 35 (16) (2001) 1417–1434.
[68] R. Chandra, S.P. Singh, K. Gupta, in: A Study on Damping Evaluation in Fiber-Reinforced Composites, India-USA Symposium on Emerging Trends in Vibration and Noise Engineering, Columbus, Ohio, The Ohio State University, December 2001.
[69] R. Chandra, S.P. Singh, K. Gupta, Micromechanical Damping Models for Fiber Reinforced Composites: A Comparative Study, Composite: Part A 33 (2002) 789–796.
[70] R.D. Adams, M.R. Maheri, Damping in Advanced Polymer–Matrix Composites, Journal of Alloys and Compounds 355 (2003) 126–130.
[71] A.K. Kaw, G.H. Besterfield, Effect of Interphase on Mechanical Behavior of Composites, Journal of Engineering Mechanics 117 (11) (1991) 2641–2658.
[72] K. De, J.M.M. Kilinken, A.A.J.N. Jandpeijs, Influence of Fiber Surface Treatment on Transverse Properties of Carbon Fiber Reinforced Composites, in: Proceedings of International Conference on Advanced Composites Materials, Published by Minerals, Metals and Materials Society. (TMS), Warrendale, PA. USA, 1993, pp. 427–432.
[73] Y.C. Chu, S.I. Rokhlin, Determination of Fiber-Matrix Interphase Moduli from Experimental Moduli of Composites with Multilayered Fibers, Mechanics of Materials 21 (3) (1995) 191–215.
[74] S.D. Gardner, C.U. Pittman Jr., R.M. Hackett, Polymeric Composite Materials Incorporating an Elastomeric Interphase, Composite Science and Technology 46 (4) (1993) 307–318.

[75] C. Haoran, S.X. Feng, F.M. Williams, Effect of Interphase on Overall Average Mechanical Properties and Local Stress Field of Multiphase Medium Materials, Computational Material Science 4 (2) (1995) 117–124.
[76] B.Y. Low, S.D. Gardner, C.U. Pittman Jr., R.M. Hackett, Micromechanical Characterization of Residual Thermal Stresses in Carbon Fiber/Epoxy Composites Containing Non-Uniform Interphase Region, Composite Engineering 5 (4) (1995) 375–396.
[77] W. Huang, S.I. Rokhlin, Generalized Self-Consistent Model for Composites with Functionally Graded and Multilayered Interphase Transfer Matrix Approach, Mechanics of Materials 22 (3) (1996) 219–247.
[78] K. Jayaraman, Z Gao, K.L Reifsnider, Interphase in Unidirectional Fiber Reinforced Epoxies: Effect on Local Stress Field, Journal of Composite Technology and Research 16 (1) (1992) 21–31.
[79] D.R. Veazie, J. Qu, Effect of Interphase on Transverse Stress-Strain Behavior in Unidirectional Fiber –Reinforced Metal Matrix Composites, Composite Engineering 5 (6) (1995) 597–610.
[80] D.S. Li, M.R. Wisnom, Micromechanical Modelling of SCS-6 Fiber-Reinforced Ti-6Al-4V Under Transverse Tension - Effect of Fiber Coating, Journal of Composite Materials 30 (5) (1996) 561–588.
[81] J.D. Achenbach, H. Zhu, Effect of Interphases on Micro and Macromechanical Behaviour of Hexagonal-Array Fiber Composites, in: Proceedings of Winter Annual Meeting Conference, ASME Publication, 1990, p. 18.
[82] S. Yi, G.D. Pollock, M.F. Ahmad, H.H. Hilton, Thermo-Viscoelastic Analysis of Fiber-Matrix Interphase, in: 34th AIAA/ASME/ASCE/AHS/ASC Structures, Structural Dynamic and Materials Conference, Part 3, 1993, pp. 1807–1815.
[83] D.D. Robertson, S. Mall, Fiber-Matrix Effect Upon Transverse Behavior of Metal-Metal- Matrix Composite, Journal of Composite Technology and Research 14 (1) (1992) 311.
[84] K. Jayaraman, K.L. Reifsnider, L. Kenneth, R.E. Swain, Elastic and Thermal Effects in the Interphase: Part I. Comments on Characterization Methods, Journal of Composite Technology and Research 15 (1) (1993) 3–13.
[85] N.R. Scotts, R.L. Cullough, Interphase in Polymer Matrix Composites, Flight-Vehicle Materials, Structures and Dynamics Assessment and Future Directions 2 (1994) 328–350.
[86] J.J. Lesko, K. Jayaraman, K.L. Reifsnider, Gradient Interphase Region in Composite System, Key Engineering Materials 116-117 (1996) 61–86.
[87] J.G. Williams, M.R. James, W.L. Morris, Formation of Interphase in Organic Matrix Composite, Composites 25 (7) (1994) 755–762.
[88] F.T. Fisher, L.C. Brinson, Viscoelastic Interphases in Polymer–Matrix Composites: Theoretical Models and Finite-Element Analysis, Composites Science and Technology 61 (2001) 731–748.
[89] T.C. Lim, Young's Modulus of Coated Inclusion Composites by The Generalized Mechanics-of-Materials (GMM) Approach, Journal of Thermoplastic Composite Materials 16 (2003) 385–402.
[90] F. Xun, G. Hu, Z. Huang, Influence of Gradual Interphase on Overall Elastic and Viscoelastic Properties of Particulate Composites, Journal of Thermoplastic Composite Materials 17 (2004) 411–426.
[91] J. Wang, L.C. Steven, G.M. Sonia, Numerical Modeling of the Elastic Behavior of Fiber-Reinforced Composites with Inhomogeneous Interphase, Composites Science and Technology 66 (2006) 1–18.
[92] L.E. Nielson, Mechanical Properties of Polymer and Composites, 2, Dekker, New York, 1974.

[93] J. Kubat, M Rigdahl, M Welander, Characterization of Interfacial Interactions in High Density Polyethylene Filled with Glass Spheres Using Dynamic-Mechanical Analysis, Journal of Applied Polymer Science 39 (7) (1990) 1527–1539.
[94] K.D. Ziegel, A. Ramanov, Modulus Reinforcement in Elastomer Composites I. Inorganic Fillers, Journal of Applied Polymer Science 17 (1973) 1119–1131l.
[95] T. Murayama, Dynamic Mechanical Analysis of Polymer Materials, Elsevier, Amsterdam, 1978.
[96] J. Chinquin, B. Chabert, J. Chauchard, E. Morel, J.P. Totignon, Characterization of Thermoplastic (Polyamide) Reinforced with Unidirectional Glass Fibers, Matrix Additives and Fibers Surface Treatment Influence on Mechanical and Viscoelastic Properties, Composites 21 (1990) 141–147.
[97] S.K. Chaturvedi, G.Y. Tzeng, Micromechanical Modeling of Material Damping in Discontinuous Fiber Three-phase Polymer Composite, Composite Engineering 1 (1) (1991) 49–60.
[98] T. Murayama, Dynamic Mechanical Analysis of Polymer Materials, Elsevier, Amsterdam, 1978.
[99] G.Y. Tzeng, Micromechanical Modeling of Damping and Stiffness of Three Phase Polymer Composite, Ohio State University, 1989 M. Tech. Thesis.
[100] J.A. Vantomme, Parametric Study of Material Damping in Fiber Reinforced Plastics, Composites 26 (1995) 147–153.
[101] I.C. Finegan, R.F. Gibson, Improvement of Damping at Micromechanical Level in Polymer Composite Materials under Transverse Loading by the use of Special Fiber Coatings, Journal of Vibration and Acoustics 120 (1998) 623–627.
[102] I.C. Finegan, R.F. Gibson, Analytical Modeling of Damping at Micromechanical Level in Polymer Composites Reinforced with Coated Fibers, Composite Science and Technology 60 (2000) 1077–1084.
[103] J. Gu, X. Zhang, M. Gu, Effect of Interphase on the Damping Capacity of Particulate—Reinforced Metal Matrix Composites, Journal of Alloys and Compounds 381 (2004) 182–187.
[104] A.S. Nowick, B.S. Berry, A Relaxation in Crystalline Solids, Academic Press, 1972.
[105] R.L. Bagley, P.J. Torvik, Fractional Calculus- A Differential Approach to the Analysis of the Viscoelastically Damped Structures, AIAA Journal 21 (5) (1983) 741–748.
[106] D.F. Golla, P.C. Heghes, Dynamics of Viscoelastic Structures – A time domain Finite Element Formulation, Journal of Applied Mechanics 52 (1985) 897–906.
[107] G.A. Lesieture, D.C. Mingori, Direct Time Domain Finite Element Modeling of Frequency-Dependent Material Damping Augmenting Thermodynamic Fields, Journal of Guidance Control and Dynamics 13 (6) (1990) 1040–1050.
[108] G.A. Lesieture, Modeling Frequency and Temperature Dependent Longitudinal Dynamic Behavior of Linear Viscoelastic Thermorhelogically Complex Long Fiber Composites, in: Proceedings of International Mechanical Conference and Exposition, Chicago, USA, 1994, pp. 1–15.
[109] T. Pritz, Analysis of Four Parameter Fractional Derivative Model of Real Solid Materials, Journal of Sound and Vibration 195 (1) (1996) 103–115.
[110] J. Li, G.A. Weng, Orthotropic Complex Moduli of a Viscoelastic Composite Reinforcement with Aligned Elliptical Fiber, Journal of Composite Materials 30 (9) (1996) 1042–1066.
[111] W.Y. Tao, G.L. Jin, Y.T. Qing, Prediction of 3-D Effective Damping Matrix and Energy Dissipation of Viscoelastic Fiber Composites, Composite Structures 54 (2001) 49–55.
[112] M.D. Rao, Recent Applications of Viscoelastic Damping for Noise Control in Automobiles and Commercial Airplanes, Journal of Sound and Vibration 262 (2003) 457–474.

[113] R.K. Patel, B. Bhattacharya, S. Basu, A Finite Element Based Investigation on Obtaining High Material Damping over a Large Frequency Range in Viscoelastic Composites, Journal Of Sound and Vibration 303 (2007) 753–766.

[114] R.D. Adams, M.R. Maheri, Dynamic Flexural Properties of Anisotropic Fibrous Composite Beams, Composites Science & Technology 50 (4) (1994) 497–514.

[115] J.M. Kenny, M. Marchetti, Elasto Plastic Behavior of Thermoplastic Composite Laminates under Cyclic Loading, Composite Structures 32 (1995) 375–382.

[116] F. Orth, L. Hoffmann, B. Zilch, C.W. Eshrenstein, Evaluation of Composite Under Dynamic Load, Composite Structures 24 (1993) 265–272.

[117] P.R. Mantena, J.G. Vaughan, R.P. Donti, M.V. Kowsika, Influence of Process Variables on the Dynamic Characteristics of Pultruded Graphite-Epoxy Composite, Vibro-Acoustic Characterization of Materials and Structures 14 (1992) 147–154 ASME Publication.

[118] P.R. Mantena, J.G. Vaughan, K. Murthy, Ultrasonic and Vibration Methods for the Characterization of Pultruted Composite, Composite Engineering 5 (12) (1995) 1443–1451.

[119] D.J. Nelson, J.W. Hancock, Interfacial Slip and Damping in Fiber Reinforced Composite, Journal of Mater. Science 13 (1978) 2429–2440.

[120] D.A. Saravanos, D.A. Hopkins, Effects of Delaminations on The Dynamic Characteristics of Composite Laminates: Analysis and Experiments, Journal of Sound and Vibration 195 (5) (1996) 977–993.

[121] V.K. Kinra, S. Ray, C. Zhu, R.D. Friend, S. Rawal, Measurement of Non-Linear Damping in Metal-Matrix Composites, Experimental Mechanics 37 (1) (1997) 5–10.

[122] Z. Zhang, G. Hartwig, Relation of Damping and Fatigue Damage of Unidirectional Fibre Composites, International Journal of Fatigue 24 (2002) 713–718.

[123] C. Kyriazoglou, F.J. Guild, Quantifying the Effect of Homogeneous and Localized Damage Mechanisms on the Damping Properties of Damaged GFRP and CFRP Continuous and Woven Composite Laminates-An FEA Approach, Composites: Part A 36 (2005) 367–379.

[124] S. Yi, K.S. Chian, H.H. Harry, Nonlinear Viscoelastic Finite Element Analyses of Thermosetting Polymeric Composites during Cool-Down after Curing, Journal of Composite materials 36 (2002) 3–17.

[125] V.S. Sokolinsky, H. Shen, L.S. Vaikhanski, R. Nutt, Experimental and Analytical Study of Nonlinear Bending Response of Sandwich Beams, Composite Structures 60 (2003) 219–229.

[126] G. Spathis, E. Kontou, Non-Linear Viscoplastic Behavior of Fiber Reinforced Polymer Composites, Composites Science and Technology 64 (15) (2004) 2333–2340.

[127] D.J. Pooler, L.V. Smith, Nonlinear Viscoelastic Response of Wood Plastic Composite Including Temperature Effect, Journal of Thermoplastic Composite Materials 17 (2004) 427–444.

[128] W.V. Paepegem, I. Baere, De J Degrieck, Modelling the Nonlinear Shear Stress–Strain Response of Glass Fibre-Reinforced Composites Part I: Experimental Results, Composites Science and Technology 66 (2005) 1455–1464.

[129] Pramanick, M. Sain, Nonlinear Viscoelastic Creep Prediction of HDPE–Agro-Fiber Composites, Journal of Composite Materials 40 (5) (2006) 417–431.

[130] F. Ellyin, Z. Xia, Nonlinear Viscoelastic Constitutive Model for Thermoset Polymers, Journal of Engineering Materials and Technology 128 (2006) 579–585.

[131] S. Sridharan, Nonlinear Viscoelastic Analysis of Composites using Competing Micromechanical Models, Journal of Composite Materials 40 (2006) 257–282.

[132] S.C. Han, A. Tabiei, W.T. Park, Geometrically nonlinear analysis of laminated composite thin shells using a modified first-order shear deformable element-based Lagrangian shell element, Composite Structures 82 (2008) 465–474.

CHAPTER 3

Measurement of damping

Various measures and measurements exist for damping evaluation in materials. Some of the methods were introduced in the previous chapter while discussing the models of damping. In reality, most of the measures of damping presume a model for damping, and accordingly, we try to evaluate the parameters of the presumed model. Some of the popular measures include damping ratio, loss factor, specific damping capacity, quality factor, and other dimensional parameters such as damping constant, loss modulus, etc. In this chapter, we shall discuss some enhancements in the methods of measurement of damping.

3.1 Measurement of damping from decay plot

We have already learnt that for the evaluation of damping, one can use the free vibration decay, which behaves as

$$x(t) = X_0 \cos(\omega t + \phi)\, e^{-\zeta w t} \tag{3.1}$$

Where $e^{-\zeta w t}$ is the decay component of the curve. However, we should be careful about the fact that the above relation is strictly true only for a single degree of freedom system. For a generic system such as a composite beam or plate or a built-up structure, we cannot guarantee that when we tap the structure, only the first mode will be dominating. In fact, the multimode decay curve would be obtained, which might look the one given in Fig. 3.1.

For the multimodal response, applying the log decrement formula in such a case would lead to erroneous results. For this purpose, a number of other methods exist. One such method is quickly applicable if we know the system's natural frequencies. The system natural frequencies can be obtained by conducting a Fourier transform of the above signal (FFT) either by means of a program or using an FFT analyzer. The resulting signal in the frequency domain would help in identifying precisely the different frequencies present in the signal. In the next step, we would identify the first mode frequency and generate a band pass filter around this frequency. This filter could be either hardwired, or it could be made computational using any of the commonly used programming software (such as Matlab or Labview). The original

Damping in Fiber Reinforced Composite Materials.
DOI: https://doi.org/10.1016/B978-0-323-91186-3.00001-0

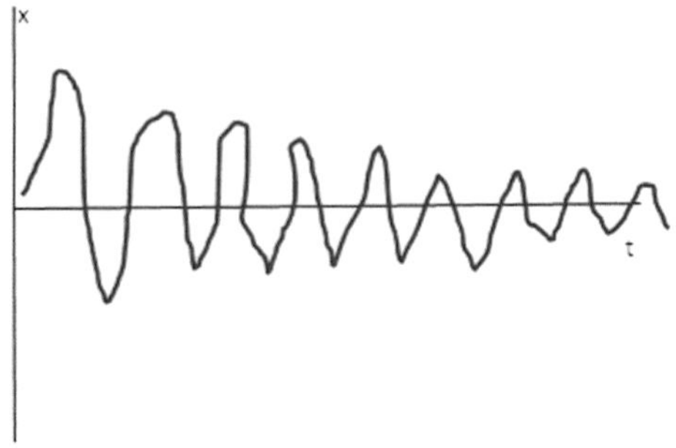

Figure 3.1 Multi-mode decay curve.

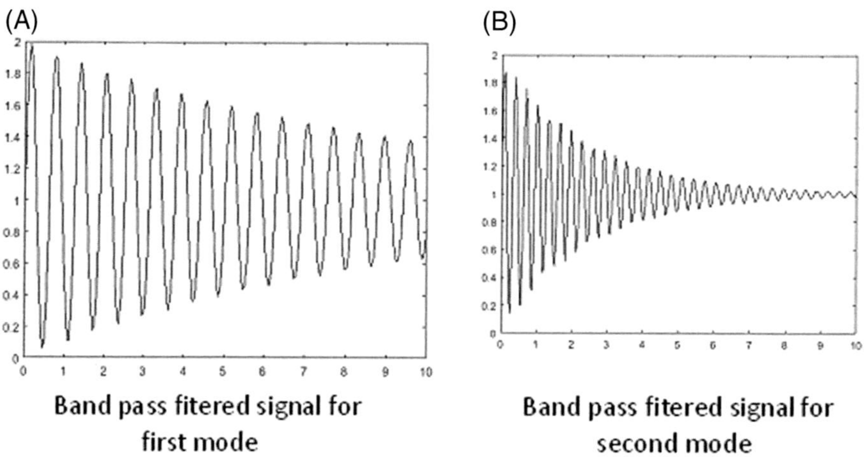

Figure 3.2 Modal damping ratio for individual modes can be obtained from the band pass filtered signal.

time-domain signal is passed through this band pass filter, and the resulting filtered signal will consist of a decay curve that will primarily correspond to the first natural frequency.

The damping obtained from the filtered curve corresponding to the first mode can also be termed as the modal damping ratio corresponding to the first mode (Fig. 3.2A). Filtering the same signal by keeping the central frequency around the second mode. We can find the modal damping ratio for the second mode as well (Fig. 3.2B).

The half-power point method described in Chapter 2 extends similarly to multimodal damping ratios. The quality factor and hence the damping ratio can also be evaluated from the frequency response function corresponding to each mode. The modal test procedures automate this process of curve fitting and also come up with better measures of the modal damping

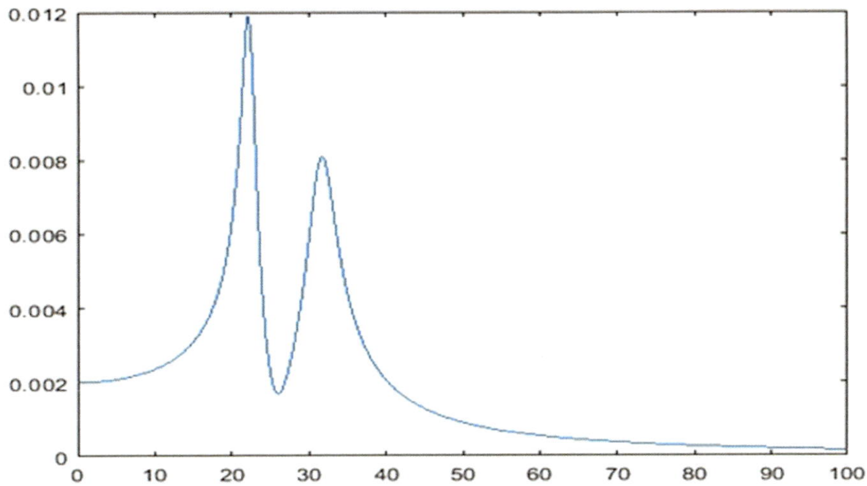

Figure 3.3 Multi modal frequency response for a typical two mode system.

ratio. These methods are less error-prone, such as the circle fit method and line fit methods [1]. Half-power gives not only accurate values of modal damping but can also indicate if the damping ratio so evaluated is frequency-dependent or not.

3.2 A generalized method of finding damping

Finding modal damping by using a traveling filter can be error prone due to incorrect tuning, and using the half power point as described in Chapter 2 can also have uncertainties because half power points are to be established at $\frac{1}{\sqrt{2}}$ times of the peak amplitude and the peak amplitude is generally never accurately known. So a number of other methods have evolved which basically rely on different types of curve fitting. These methods form the details of a separate subject called experimental modal analysis of a structure. However, we shall explain in detail one method which is termed to be more accurate, and requires lesser amount of calculation. Assume that we have obtained the frequency response function of a system in magnitude as well as phase, as shown in Fig. 3.3.

For any specific mode, let us say the second mode is the task of focus. We choose specific points before and after the resonance, and utilize magnitude as well as the phase angle (Fig. 3.4). The magnitude and phase, if plotted on an argand diagram for a particular mode, will appear as a circle. A typical such circle is shown in Fig. 3.5. In the case of limited points, a

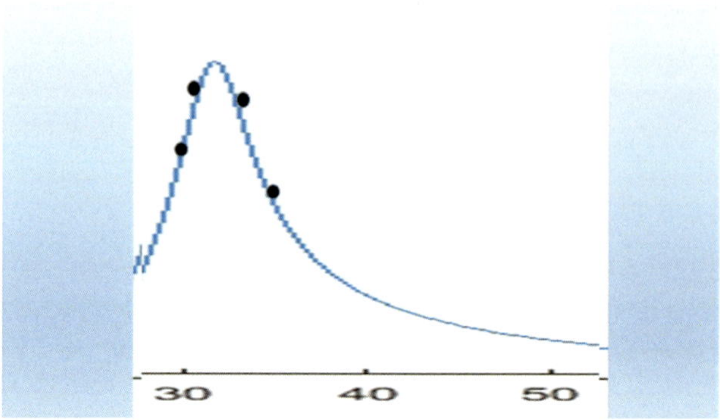

Figure 3.4 A specific mode of vibration picked for analysis.

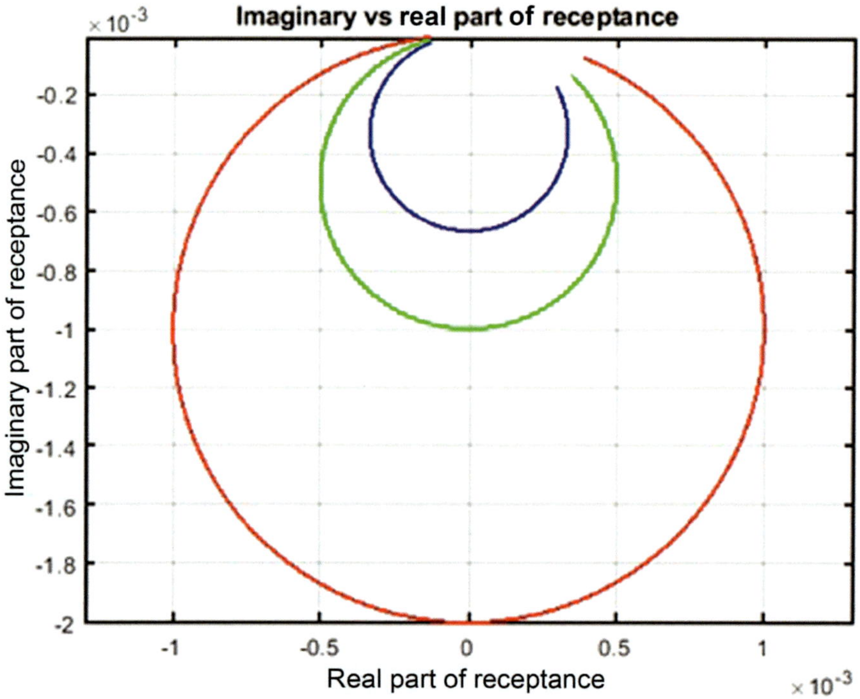

Figure 3.5 Circle plot for (argand plot of FRF).

circle fit can always be obtained. Each circle corresponds to a particular mode.

If we take a few points in the region of second mode and fit them on a phasor plot, then the plot will be as shown in Fig. 3.6. Consider two points A and B on this curve, which form an angle of θ_a and θ_b about the vertical

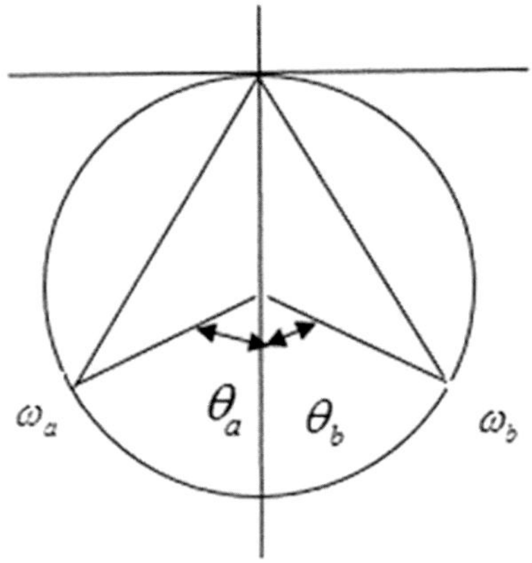

Figure 3.6 Evaluation of damping from two points on the circle.

axis and one of the two opposite sides, which means point A is preresonance and point B is postresonance. Then one can prove the exact relationship.

$$\eta_r = \frac{\omega_a^2 - \omega_b^2}{\omega_r^2\left(\tan\frac{\theta_a}{2} + \tan\frac{\theta_b}{2}\right)} \tag{3.2}$$

In case $\theta_a = 90°$ and $\theta_b = 90°$

$$\eta_r = \frac{\omega_a^2 - \omega_b^2}{\left(\frac{\omega_a + \omega_b}{2}\right) * \sqrt{2}} = \frac{\omega_b - \omega_a}{2\omega_r} \tag{3.3}$$

Then the relation becomes as the half power point method. The benefit of this method using Eq. (3.2) is that, we are not required to measure the resonance amplitude. From any two points on the resonance curve, one can get an estimate of a modal loss factor.

3.3 Multimode evaluation of damping

In the case of proportional damping, which is a commonly used assumption, one can express the damping as a function of mass and stiffness matrices. Such that

$$C = \alpha M + \beta K \tag{3.4}$$

The values of $\alpha$and β can be estimated experimentally if we know the damping of two dominant modes of the structure.

Let us suppose that, as shown in Fig. 3.2 above, the two modes have damping ratios ξ_1 and ξ_2 which has been evaluated using the methods described in the previous chapter or in the previous sections.

If ω_1 and ω_2 are the two natural frequencies of these modes, then the following relations exist.

$$C = \alpha M + \beta K$$
$$\begin{bmatrix} \ddots & & \\ & 2\xi\omega_n & \\ & & \ddots \end{bmatrix} = \alpha \begin{bmatrix} \ddots & & \\ & I & \\ & & \ddots \end{bmatrix} + \beta \begin{bmatrix} \ddots & & \\ & \omega_n^2 & \\ & & \ddots \end{bmatrix} \tag{3.5}$$

Thus

$$\begin{aligned} 2\xi_1\omega_1 &= \alpha + \beta\omega_1^2 \\ 2\xi_2\omega_2 &= \alpha + \beta\omega_2^2 \end{aligned} \tag{3.6}$$

From these two equations, one can simultaneously solve for α and β as

$$\begin{aligned} \beta(\omega_1^2 - \omega_2^2) &= 2\xi_1\omega_1 - 2\xi_2\omega_2 \\ \beta &= \frac{2\xi_1\omega_1 - 2\xi_2\omega_2}{\omega_1^2 - \omega_2^2} \end{aligned} \tag{3.7}$$

and

$$\begin{aligned} \alpha &= \frac{2\xi_1\omega_1\omega_2^2 - 2\xi_2\omega_2\omega_1^2}{\omega_2^2 - \omega_1^2} \\ &= \frac{2\omega_1\omega_2(\xi_1\omega_1 - \xi_2\omega_2)}{\omega_2^2 - \omega_1^2} \end{aligned} \tag{3.8}$$

In case more than two modes are involved in a vibration phenomenon, then one can use the matrix methods and obtain the pseudo-inverse of a rectangular matrix.

The values thus obtained will be the best least-square fitted estimate, which will result in the *least errors of prediction*. One can write Eq. (3.2) as;

$$\begin{pmatrix} 1 & \omega_1^2 \\ 1 & \omega_2^2 \\ \vdots & \vdots \\ 1 & \omega_n^2 \end{pmatrix} \begin{pmatrix} \alpha \\ \beta \end{pmatrix} = \begin{pmatrix} 2\xi_1\omega_1 \\ 2\xi_2\omega_2 \\ \vdots \\ 2\xi_n\omega_n \end{pmatrix} \tag{3.9}$$

The values of α and β obtained in this manner can be used to evaluate the damping matrix [C] which can be used in the equation of motion of the composite structure with damping.

Alternatively, when using the viscous representation, the modal loss factors can also be evaluated, and hence leading to the evaluation of the quadrature matrix "D" which for a harmonic motion follows the following:

$$\begin{aligned} &M\ddot{x} + [[K] + i[D]]x = \{f\} \\ &\text{where } D = \alpha[M] + \beta[K] \end{aligned} \tag{3.10}$$

Using orthogonal mode shapes

$$\begin{bmatrix} \ddots & & \\ & \eta_r\omega_r & \\ & & \ddots \end{bmatrix} = \alpha[I] + \beta \begin{bmatrix} \ddots & & \\ & \omega_r^2 & \\ & & \ddots \end{bmatrix} \tag{3.11}$$

Considering two modes

$$\begin{aligned} \eta_1\omega_1^2 &= \alpha + \beta\omega_1^2 \\ \eta_2\omega_2^2 &= \alpha + \beta\omega_2^2 \end{aligned} \tag{3.12}$$

From the above equation

$$\begin{aligned} \beta &= \frac{\eta_2\omega_2^2 - \eta_1\omega_1^2}{\omega_2^2 - \omega_1^2} \\ \text{and } \alpha &= \frac{\omega_1^2\omega_2^2(\eta_1 - \eta_2)}{\omega_2^2 - \omega_1^2} \end{aligned} \tag{3.13}$$

The [D] matrix evaluated using Eq. (3.10) and the obtained values of α and β can be used in the equation of motion.

3.4 Damping ratio for different modes of deformation

A fiber-reinforced material is in general an anisotropic material, and is in many cases used as an orthotropic material conformation. It has different values of moduli in three principal directions as well as in the shear mode of deformation. These differences arise in composite materials primarily due to different levels of contributions to the strain energy by the individual constituents, viz. the fiber and the matrix, at the micromechanics level. Extending the same argument to a dynamic case, the fiber reinforced composite is likely to have different damping ratios or loss factors in different directions. In the case of a hysteretic damping mechanism, this extension is simplified for a harmonic loading since the stiffness matrix [K] is simply transformed into its complex form [K]+I[D] where the [D] matrix is the matrix of loss moduli corresponding to each stiffness coefficient.

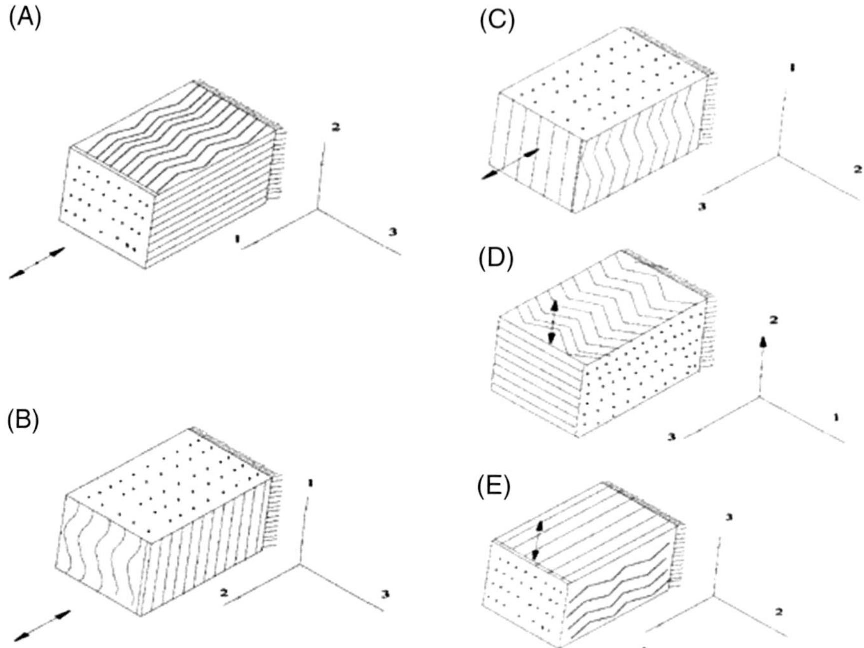

Figure 3.7 Various modes of Deformation of a cuboid with unidirectional fiber reinforcement (A) 11 mode, (B) 22 mode, (C) 33 mode, (D) 13 mode, and (E) 23 mode (Chandra et al. [2]).

For a simplified, linear viscoelastic material, the ration of any element of D say d_{ij} to the corresponding stiffness matrix term, say k_{ij} is constant and is termed as the loss factor η_{ij}. Thus, the aim of the experimental procedures becomes to target and find the loss factors in various deformation modes.

One of the initial attempts in evaluation of different damping coefficients of a composite specimen has been made by Chandra et al. [2]. They experimented on a cuboid specimen and a circular tube to evaluate the damping coefficient (loss factor) in different modes. The representative values of η_{11}, η_{22}, tη_{33}, η_{13} and η_{23} are determined by a crude method of ramp test on a rectangular block containing unidirectional fiber, while the in plane shear loss factor η_{12} was determined by the torsional vibrations of a unidirectional hollow bar. The various modes of deformation are given in Fig. 3.7.

The 12 modes which involved inplane shear was evaluated and separated using the torsional vibration of a unidirectional bar. In order to validate the results thus obtained, bending vibrations from various beams made of unidirectional composite beams were generated with fibers aligned in different directions. In bending, the contribution to bending strain energy comes from stretching along the axis as well as thickness shear acting. The equivalent loss factor is determined from the resultant total energy dissipated and the total strain energy. The measured loss factor matrix thus obtained for the various deformation modes is given as under (for a typical case of fiber reinforced composite).

$$\begin{bmatrix} \eta_{11} & \eta_{12} & \eta_{13} \\ \eta_{21} & \eta_{11} & \eta_{23} \\ \eta_{31} & \eta_{32} & \eta_{33} \end{bmatrix} = \begin{bmatrix} 252 & 197.3 & 255 \\ 197.3 & 257 & 255 \\ 255 & 255 & 256 \end{bmatrix} \times 10^{-4} \tag{3.14}$$

These values, though in the range of values calculated analytically, are on the higher side, because by tapping a cuboid we cannot ensure a pure tensile or pure shear deformation mode as is needed.

A more consistent method of evaluating the anisotropic loss factor matrix is proposed by [5] and is described in the next section.

3.5 Resonalyser method of evaluation of damping coefficient matrix

In the resonalyser method, the damping matrix and the stiffness matrix could be identified for a composite specimen. The method essentially uses different modes of vibration and their natural frequencies to tune in the orthotropic material constants for a single orthotropic lamina. The relationship between stress and strains is given as under

$$\begin{Bmatrix} \varepsilon_1 \\ \varepsilon_2 \\ \gamma_{12} \end{Bmatrix} = \begin{bmatrix} \frac{1}{E_1} & \frac{-\nu_{12}}{E_1} & 0 \\ \frac{-\nu_{12}}{E_1} & \frac{1}{E_2} & 0 \\ 0 & 0 & \frac{1}{G_{12}} \end{bmatrix} \begin{Bmatrix} \sigma_1 \\ \sigma_2 \\ \tau_{12} \end{Bmatrix} \tag{3.15}$$

The five engineering constants in terms of principle directions are the properties to be determined. Soh and Oomens [3] used thin rectangular plates modeled as Kirchoff plates to identify the material stiffness. One can use a set of natural frequencies or mode shapes to define an objective function as the difference between the actual parameters obtained from the experiments and set up minimization problems using the above unknown constants as the design variables. Still, to ensure that the optimization leads to the correct values one should choose the initial values of the parameters judiciously. In that manner the procedures are similar to the model updating methods described in the standard texts on modal analysis. Additional as an extension, one can identify the damping parameters of the composite materials, in fact the complete identification is better done in two stages first the stiffness identification using the natural frequencies and mode shapes and secondly the complete identification using the frequency response functions.

3.6 Frequency dependence of damping

The general a frequency dependence damping a moduli of a material can be expressed as a series

$$a_0\sigma + a_1 D\sigma + a_2 D^2\sigma + ...a_m D^m\sigma = b_0\varepsilon + b_1 D\varepsilon + b_2 D^2\varepsilon + ...b_n D^n\varepsilon$$
where D is the derivative operator $\frac{d}{dt}$
(3.16)

If one ignores all the rate dependent terms, the above equation is statement of Hooke's law with b_0/a_0 being the Young's modulus. In general, the frequency dependence of damping coefficient or loss factor is modelled by taking a number of terms in the above equation. Different models have been proposed, with varying degrees of complexity. Generally, a three-parameter model or four-parameter model is simple to implement. But one can go in for a more precise representation using a seven parameter model or higher [4]. The coefficients have to be determined by the curve fitting procedure on the experimentally measured data. One more important point to understand is the measurement of frequency dependence. This is not established by impulse test or resonance curve fitting since the resonance would occur at specific frequencies. Rather the forced vibration test is used in which both stress and strain variation is measured as a function of time. The phase difference between stress wave and strain wave (strain lags) is indicative of the prevalent damping at that frequency. This procedure has been automated in highly precise machines called dynamic mechanical analyser (DMA).

3.7 Concluding remarks

Determination of damping coefficients in composite materials has been a formidable challenge. While one can easily conduct experiments and calculate the damping ratios or loss factors for different modes of vibration or even along the geometrical axes of the specimen. But to get the damping coefficients to be used in conjunction with the modulus values is a complex task. This is because the composite materials have vastly different properties along the materials axes viz. along with the fiber and across the fiber directions and different values exist along with the shear-associated moduli.

For obtaining approximate models of the composite structures one can use as a good approximation the assumption of proportional damping. The damping coefficients or the loss factors coefficients α and β can be found using the modal damping ratios of two modes of interest. In that evaluation, a least square formulation can be evolved as has been demonstrated in this chapter in order to get the values of proportionality constants which can give an overall good validation of the frequency range of interest.

The circle fit method of determination of modal loss factor takes into account the limitation of not accurately estimating the resonance frequency and it has been demonstrated that one need not have the exact resonance point but two or three points around the resonance can be used to get a better estimated of damping.

References

[1] D.J. Ewins, Modal Testing: Theory and Practice, Research Studies Press Ltd, 16 Coach House Cloisters, 10 Hitchin Street, Baldock, Hertfordshire, England, SG7 6AE, 2009.

[2] R. Chandra, S.P. Singh, K. Gupta, Experimental evaluation of damping of fiber reinforced composites, J. Compos. Technol. Res. ASTM, (2016) 1–9.

[3] H. Soh, C. Oomens, Material Identification Using Mixed Numerical Experimental Methods, Kluwer Academic Publishers, PO Box 17, 3300 AA, Dordrecht, The Netherlands, 1997.

[4] B.J. Lazan, Damping of Materials and Members in Structural Mechanics, Elsevier Science and Technology, Amsterdam, Netherlands, 1968.

[5] T.L. Auwagic, H. Soi, R. Roebben, W. Heylen, Y Shi, Validation of the resonalyser method: an inverse method for material identification, in: Proceedings of ISMA, Vol.II, 2002, pp. 687–694.

CHAPTER 4

Micromechanical study of two-phase composite

Micromechanics applies theories of elasticity and plasticity to study imperfections in crystals and inclusions and inhomogeneities in composite materials. Micromechanical modeling has been demonstrated to be a powerful tool to predict macroscopic constitutive response, especially for those materials having a periodic microstructure. In micromechanics, materials are represented at the microscale by mathematical models which explicitly incorporate features of the material microstructure such as second phase or reinforcing particles, voids, defects, microcracks, grain boundaries, and crystal structure. Modeling at this level gives great insight into the real physical processes that occur in materials, which dictate the overall mechanical performance of engineering and industrial components. Micromechanical improvements in composite material damping result from changes in damping properties and geometries at or below the lamina constituent level. A micromechanical approach has been used for both continuous and short fiber-reinforced composite materials. Some important micromechanical models, with varying degrees of complexities used in the literature for estimation of elastic moduli are adopted to predict damping coefficients for continuous fiber-reinforced composites.

4.1 Micromechanical models

Micromechanics deals with the interaction between fiber and matrix to evaluate the mechanical performance of a composite. Some of the important micromechanical models are discussed here for the prediction of elastic moduli and loss factors.

4.1.1 Hashin model

The elastic moduli are predicted by Hashin [1] for fiber composites with anisotropic constituents. Here the constituents are considered to be isotropic and modified equations for the elastic moduli are expressed in terms of complex form using the correspondence principle. The longitudinal

Damping in Fiber Reinforced Composite Materials.
DOI: https://doi.org/10.1016/B978-0-323-91186-3.00007-1

modulus E_{11} in extension and transverse modulus E_{22} in complex form are given as:

$$E_{11}^{*} = E_{11}^{*} = E_{f}^{*}V_{f} + E_{m}^{*}V_{m} + \frac{4\left(\nu_{f} - \nu_{m}\right)^{2}V_{f}V_{m}}{\frac{V_{m}}{k_{f}^{*}} + \frac{V_{f}}{k_{m}^{*}} + \frac{1}{G_{m}^{*}}} \tag{4.1}$$

$$E_{22}^{*} = \frac{4k^{*}G_{23}^{*}}{k^{*} - m^{*}G_{23}^{*}} \tag{4.2}$$

Further, transverse shear modulus in complex form is expressed as:

$$G_{23}^{*} = G_{m}^{*} + \frac{V_{f}}{\frac{1}{G_{f}^{*} - G_{m}^{*}} + \frac{\left(k\rightarrow_{m}^{*} + 2G_{m}^{*}\right)}{2G_{m}^{*}\left(k_{m}^{*} - 2G_{m}^{*}\right)}V_{m}} \tag{4.3}$$

Complex longitudinal shear modulus is obtained as

$$G_{12}^{*} = G_{m}^{*} + \frac{V_{f}}{\frac{1}{G_{f}^{*} - G_{m}^{*}} + \frac{V_{m}}{2G_{m}^{*}}} \tag{4.4}$$

where k^{*} is bulk modulus of composite in complex form and can be written in term of the moduli of constituents as follows,

$$k^{*} = k_{m}^{*} + \frac{V_{f}}{\frac{1}{k_{f}^{*} - k_{m}^{*}} + \frac{V_{m}}{k_{m}^{*} + G_{m}^{*}}} \tag{4.5}$$

The Poisson's ratio is given by,

$$\nu_{12} = \nu_{f}V_{f} + \nu_{m}V_{m} + \frac{\left(\nu_{f} - \nu_{m}\right)\left(\frac{1}{k_{m}^{*}} - \frac{1}{k_{f}^{*}}\right)V_{f}V_{m}}{\frac{V_{f}}{k_{m}^{*}} + \frac{V_{m}}{k_{f}^{*}} + \frac{1}{G_{m}^{*}}} \tag{4.6}$$

Loss factors are obtained from Hashin's model as the ratio of loss to storage moduli. In the above equations, symbol E stands for Young's modulus, G for shear modulus, and K for bulk modulus. Subscript * means the quantity is in complex form, including its loss modulus. The subscripts f and m refer to fiber and matrix. Subscripts 1, 2, and 3 refer to material axes.

4.1.2 Unified micromechanics

Unified micromechanics applied to single fiber representative volume element (RVE) of square packing array based on energy method is proposed by Saravanos-Chamis [2]. This approach overcomes the restriction on the isotropic elastic and dissipative properties of fiber and matrix posed by earlier theories.

The loss factor in the longitudinal direction is derived on the basis of the rule of mixtures, assuming uniform strain across the section. The longitudinal stress in the fiber and matrix is of different level, but uniform within each constituent. The loss factor in longitudinal direction 1 is given as,

$$\eta_{11} = \eta_{11}^{f} V_f \frac{E_{11}^{f}}{E_{11}} + \eta_m V_m \frac{E_m}{E_{11}} \tag{4.7a}$$

Where;

$$E_{11} = E_{11}^{f} V_f + E_m V_m \tag{4.7b}$$

Similarly, transverse normal loss factor is expressed as,

$$\eta_{22} = \eta_{22}^{f} \sqrt{V_f} \frac{E_{22}}{E_{22}^{f}} + \eta_m \left(1 - \sqrt{V_f}\right) \frac{E_{22}}{E_m} \tag{4.8}$$

where,

$$E_{22} = \left(1 - \sqrt{V_f}\right) E_m + \frac{\sqrt{V_f} E_m}{1 - \sqrt{V_m \left(1 - \frac{E_m}{E_{22}^{f}}\right)}} \tag{4.9}$$

Here, the damping of fiber and matrix is assumed to be isotropic in nature. Transverse isotropy of the fibers and consequently of the composite leads to the following equality for the transverse through-thickness damping,

$$\eta_{33} = \eta_{22} \tag{4.10}$$

Further, in plane shear damping η_{12} is obtained as,

$$\eta_{12} = \eta_{12}^{f} \sqrt{V_f} \frac{G_{12}}{G_{12}^{f}} + \eta_m \left(1 - \sqrt{V_f}\right) \frac{G_{12}}{G_m} \tag{4.11}$$

where

$$G_{12} = \left(1 - \sqrt{V_f}\right) G_m + \frac{\sqrt{V_f} G_m}{\left\{1 - \sqrt{V_f} \left(1 - \frac{G_m}{G_{12}^{f}}\right)\right\}} \tag{4.12}$$

Through the thickness shear damping or the interlaminar shear damping η_{13} is equal to the inplane shear damping due to transverse isotropy. Therefore

$$\eta_{13} = \eta_{12} \tag{4.13}$$

And the transverse shear loss factor is expressed as

$$\eta_{23} = \eta_{f23}\sqrt{V_f}\frac{G_{23}}{G_{23}^f} + \eta_m\left(1 - \sqrt{V_f}\right)\frac{G_{23}}{G_m} \tag{4.14}$$

where

$$G_{23} = \frac{E_{22}}{2(1 + \nu_{23})}$$

and

$$\nu_{23} = \frac{\nu_m}{\left(1 - V_f \nu_m\right)} + V_f\left\{\nu_{23}^f - \frac{\left(1 - V_f\right)\nu_m}{\left(1 - V_f \nu_m\right)}\right\} \tag{4.15}$$

4.1.3 Eshelby's method

Based on the stress distribution in inclusions, Eshelby (1957) gave the relation for various moduli. These are extended to complex form [17] to give the damping coefficients.

Longitudinal modulus E_{11} in pure tension in complex form is given as

$$E_{11}^* = \frac{E_m^*}{1 + V_f\left[A_1^* - \nu_m\left(A_2^* + A_3^*\right)\right]/A^*} \tag{4.16}$$

Transverse modulus E_{22} in complex form is given as

$$E_{22}^* = \frac{E_m^*}{1 + V_f\left[A_5^* - \nu_m\left(A_4^* + A_6^*\right)\right]/A^*} \tag{4.17}$$

Longitudinal shear modulus G_{12} is expressed in complex form as

$$G_{12}^* = \left[1 + \frac{V_f}{2V_m S_{1212} + G_m^*/\left(G_f^* - G_m^*\right)}\right]G_m^* \tag{4.18}$$

Plane shear modulus G_{23} is expressed in complex form as

$$G_{23}^* = \left[1 + \frac{V_f}{2V_m S_{2323} + G_m^*/\left(G_f^* - G_m^*\right)}\right]G_m^* \tag{4.19}$$

Four elastic moduli expressed in complex form are used to predict the damping coefficients as ratio of loss and storage moduli. This formulation incorporates the effect of microgeometry and the fiber volume fraction.

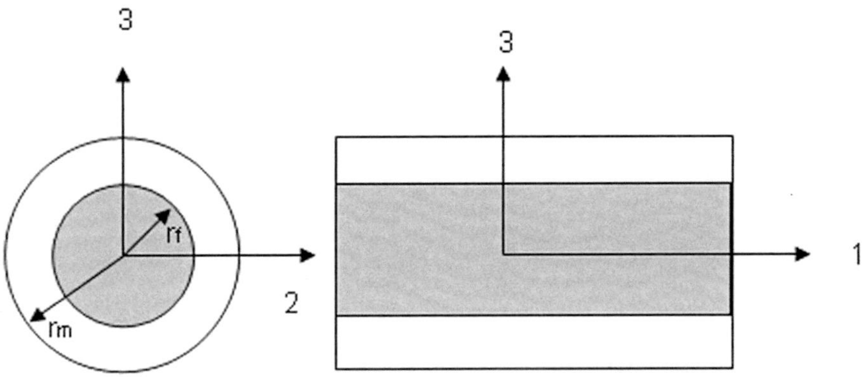

Figure 4.1 Concentric cylinder model.

4.1.4 Bridging model

Huang's bridging model [3] for unidirectional fiber-reinforced composite has been discussed. In this model, elastic stresses in matrix material are correlated with the fiber by bridging matrix. The bridging matrix represents the load sharing capacity of one constituent phase in the composite with respect to the other phase.

$$\left\{d\sigma_i^m\right\} = \left|A_{ij}^{mf}\right|\left\{d\sigma_i^f\right\} \tag{4.20}$$

Where

$$\{d\sigma_i\} = \{d\sigma_{11}, d\sigma_{22},\ d\sigma_{33},\ d\sigma_{13},\ d\sigma_{12}, d\sigma_{23}\ \}^T$$

$[A_{ij}]$ = bridging matrix

Suffix f and m refer fiber and matrix respectively, whereas quantity without suffix represents the composite.

Concentric cylinder model for unidirectional two-phase fiber-reinforced composite is shown in Fig. 4.1 in the form of RVE. It consists of fiber of radius $\mathrm{r_f}$ embedded into matrix of radius $\mathrm{r_m}$. Further condition of perfect fiber-matrix bonding and variable state of stress is assumed within RVE. Formulation has been done for the elastic deformation.

Let V' denotes the volume of RVE of unidirectional fiber-reinforced composite as shown in Fig. 4.1, correspondingly V'_f and V'_m represents volume of fiber and matrix respectively.

Suppose that i th point represents different stress at different points in RVE. Volume average stress $\overline{\sigma_i}$ of the composite is defined by Eq. (4.21).

$$\overline{\sigma_i} = \frac{1}{V'} \int \sigma_i dv = \frac{1}{V'} \left[\int_{v'_f} \sigma_i dv + \int_{v'_m} \sigma_i dv \right] \tag{4.21}$$

$$\overline{\sigma_i} = \left[\left(\frac{V'_f}{V'} \right) \left(\frac{1}{V'_f} \int_{v'_f} \sigma_i dv \right) + \left(\frac{V'_m}{V'} \right) \left(\frac{1}{V'_m} \int_{v'_m} \sigma_i dv \right) \right], \ \overline{\sigma_i} = V_f \overline{\sigma_i^f} + V_m \overline{\sigma_i^m} \tag{4.22}$$

where $i = 1$ to 6.

$$\overline{\sigma_i^f} = \left(\frac{1}{V'_f} \int_{v'_f} \sigma_i dv \right), \ \overline{\sigma_i^m} = \left(\frac{1}{V'_m} \int_{v'_m} \sigma_i dv \right) \tag{4.23}$$

$$V_f = \frac{V'_f}{V'}, \ V_m = \frac{V'_m}{V'}$$

where V_f and V_m volume fraction of fiber and matrix respectively and $\overline{\sigma_i^f}$ and $\overline{\sigma_i^m}$ are volume averaged constituent stresses. Similarly, for volume averaged strain is given by Eq. (4.24),

$$\overline{\varepsilon_i} = V_f \overline{\varepsilon_i^f} + V_m \overline{\varepsilon_i^m} \tag{4.24}$$

In further discussion volume averaged stresses and strain, will be represented by omitting the over bar.

Volume average stressed and strains in RVE of lamina in incremental form is given by Eqs. (4.25) and (4.26)

$$\{d\sigma\} = V_f \{d\sigma^f\} + V_m \{d\sigma^m\} \tag{4.25}$$

And

$$\{d\varepsilon\} = V_f \{d\varepsilon^f\} + V_m \{d\varepsilon^m\} \tag{4.26}$$

Constitutive equations correlating the averaged stresses and strains in different phases of the RVE are expressed by Eqs. (4.27) to (4.29)

$$\{d\varepsilon^f\} = |S^f| \{d\sigma^f\} \tag{4.27}$$

$$\{d\varepsilon^m\} = |S^m| \{d\sigma^m\} \tag{4.28}$$

Substituting Eqs. (4.27) and (4.28) into Eq. (4.26)

$$\{d\varepsilon\} = V_f |S^f| \{d\sigma^f\} + V_m |S^m| \{d\sigma^m\} \tag{4.29}$$

Using Eqs. (4.20) and (4.21) into Eq. (4.29)

$$\{d\varepsilon\} = V_f\left|S_{ij}^f\right|\{d\sigma^f\} + V_m|S^m|\left|A_{ij}^{mf}\right|\{d\sigma^f\}$$

$$\{d\varepsilon\} = \left[V_f\left|S_{ij}^f\right| + V_m\left|S_{ij}^m\right|\left|A_{ij}^{mf}\right|\right]\{d\sigma^f\} \tag{4.30}$$

$$\{d\varepsilon\} = \left|S_{ij}\right|\{d\sigma\} = \left|S_{ij}\right|\left[V_f + V_m\left|A_{ij}^{mf}\right|\right]\{d\sigma^f\}$$

$$S_{ij} = \left[V_f\left|S_{ij}^f\right| + V_m\left|S_{ij}^m\right|\left|A_{ij}^{mf}\right|\right]\left[V_f[I] + V_m\left|A_{ij}^{mf}\right|\right]^{-1} \tag{4.31}$$

where [I] = unit matrix.

Overall compliance matrix of lamina is given by Eq. (4.32)

$$S_{ij} = \begin{bmatrix} [S_{ij}]_\sigma & 0 \\ 0 & [S_{ij}]_\tau \end{bmatrix} \tag{4.32}$$

$$[S_{ij}]_\sigma = \begin{bmatrix} 1/E_{11} & -\nu_{12}/E_{11} & -\nu_{12}/E_{11} \\ & 1/E_{22} & -\nu_{23}/E_{22} \\ symmetry & & 1/E_{22} \end{bmatrix} \tag{4.32a}$$

$$[S_{ij}]_\tau = \begin{bmatrix} 1/G_{23} & 0 & 0 \\ 0 & 1/G_{12} & 0 \\ 0 & 0 & 1/G_{12} \end{bmatrix} \tag{4.32b}$$

Where $[S_{ij}]_\sigma$ and $[S_{ij}]_\tau$ are normal and shear compliance matrix.

Material parameters E_{22}, G_{23}, and ν_{23} are related to each other as in Eq. (4.33)

$$G_{23} = \frac{E_{22}}{2(1+\nu_{23})} \tag{4.33}$$

Therefore, only five independent elastic moduli exist; hence only five independent elements in the bridging matrix. General form of bridging matrix is given by Eq. (4.34).

$$[A_{ij}] = \begin{bmatrix} a_{11} & a_{12} & a_{13} & 0 & 0 & 0 \\ a_{21} & a_{22} & a_{23} & 0 & 0 & 0 \\ a_{31} & a_{32} & a_{33} & 0 & 0 & 0 \\ 0 & 0 & 0 & a_{44} & 0 & 0 \\ 0 & 0 & 0 & 0 & a_{55} & 0 \\ 0 & 0 & 0 & 0 & 0 & a_{66} \end{bmatrix} \tag{4.34}$$

The parameters $a_{11}, a_{22}, a_{33}, and\ a_{23}$ are considered to be independent.

Bridging matrix between fiber and matrix is represented by Eq. (4.35)

$$\left[A_{ij}^{fm}\right] = \begin{bmatrix} a_{11}^{fm} & a_{12}^{fm} & a_{13}^{fm} & 0 & 0 & 0 \\ a_{21}^{fm} & a_{22}^{fm} & a_{23}^{fm} & 0 & 0 & 0 \\ a_{31}^{fm} & a_{32}^{fm} & a_{33}^{fm} & 0 & 0 & 0 \\ 0 & 0 & 0 & a_{44}^{fm} & 0 & 0 \\ 0 & 0 & 0 & 0 & a_{55}^{fm} & 0 \\ 0 & 0 & 0 & 0 & 0 & a_{66}^{fm} \end{bmatrix} \quad (4.35)$$

The independent elements a_{11}, a_{22}, a_{33}, and a_{23} depends on the elastic properties of the matrix and fiber and also on fiber packing.

Due to the axisymmetry of concentric cylinder

$$a_{33} = a_{22} \quad (4.36a)$$

$$a_{13}^{fm} = a_{12}^{fm} = \frac{\left(S_{12}^{f} - S_{12}^{m}\right)\left(a_{11}^{fm} - a_{22}^{fm}\right)}{\left(S_{11}^{f} - S_{11}^{m}\right)} \quad (4.36b)$$

$$a_{11}^{fm} = \frac{E^{m}}{E^{f}} \quad (4.36c)$$

$$a_{22}^{fm} = \beta + (1-\beta)\frac{E^{m}}{E^{f}} \quad (4.36d)$$

$$a_{44}^{fm} = \beta + (1-\beta)E^{m}/E^{f} \quad (4.36d)$$

$$a_{55}^{fm} = a_{66}^{fm} = \alpha + (1-\alpha)G^{m}/G^{f} \quad (4.36e)$$

where α and β are fiber packing factors.

Using the bridging matrices, four elastic moduli E_{11}, E_{22}, G_{12}, and G_{23} of unidirectional fiber composite are derived and given by Eq. (4.37) to Eq. (4.40) respectively.

$$E_{11} = V_f E_{11}^{f} + V_m E_{11}^{m} \quad (4.37)$$

$$E_{22} = \frac{\left(V_f + V_m a_{11}^{fm}\right)\left(V_f + V_m a_{22}^{fm}\right)}{\left(V_f + V_m a_{11}^{fm}\right)\left(V_f S_{22}^{f} + V_m a_{22}^{fm} S_{22}^{m}\right) + \left\{V_f V_m \left(S_{21}^{m} - S_{21}^{f}\right) a_{12}^{fm}\right\}} \quad (4.38)$$

$$G_{12} = \frac{\left(V_f + V_m a_{66}^{fm}\right)\left(G_f G_m\right)}{\left(V_f G_m + V_m a_{66}^{fm} G_f\right)} \quad (4.39)$$

$$G_{23} = \frac{0.5\left(V_f + V_m a_{44}^{fm}\right)}{V_f\left(S_{22}^{f} - S_{23}^{f}\right) + V_m a_{44}^{fm}\left(S_{22}^{m} - S_{23}^{m}\right)} \quad (4.40)$$

Further, applying the correspondence principle to Eq. (4.37) to Eq. (4.40), loss factors are predicted and its variation is demonstrated with respect to fiber volume fraction.

4.2 Fiber packing geometry

The incorporation of fiber packing geometry in fiber-reinforced composite determines the level of interaction between the fibers and its effect on effective elastic and dynamic mechanical properties. In the following section, some important fiber packing geometries are discussed and some selected ones are used in FEM/strain energy modeling of fiber-reinforced composite for the prediction of damping.

4.2.1 Types of fiber packing

The fiber distribution in the transverse plane has been generally idealized for both analytical and numerical methods through simple, square, and hexagonal patterns of packing geometry. In all such approaches, uniform distribution and periodic packing of fibers is assumed (Figs. 4.2A, B, and C), which simplifies the numerical modeling because of the RVE of limited size and simple boundary conditions. Proper selection of fiber packing geometry leads to a better understanding of fiber matrix interaction and its effect on damping in composites. Foey [4] was the first to adopt three simple cell models (square and hexagonal) and FEM to estimate Poisson's ratios, elastic moduli, and stress distribution around a fiber. [18] for a rectangular array of fibers and Achenbach and Zhu [5] for hexagonal array of fibers presented the studies to include the effect of interphase. Several other investigations addressed to the behavior of unidirectional polymers as well as metal matrix composites, restricted to the use of a regular type of fiber packing geometry. Zhang et al. [6] presented two types of local non-uniform fiber packing geometries accounting for inhomogeneous fiber distribution, retaining the convenience of RVE. Jungki et al. (2006) [19] analyzed the stress field at the interface between the matrix and the central orthotropic inclusion for square, hexagonal, and random packing of the inclusions using volume integral equation method (VIEM). Wongsto and Li [7] developed the micromechanical finite element analysis for unidirectionally fiber-reinforced composites having fibers distributed at random over the transverse cross-section and predicted transverse Young's modulus and also compared them with those by assuming regular packing in relation to experimental data.

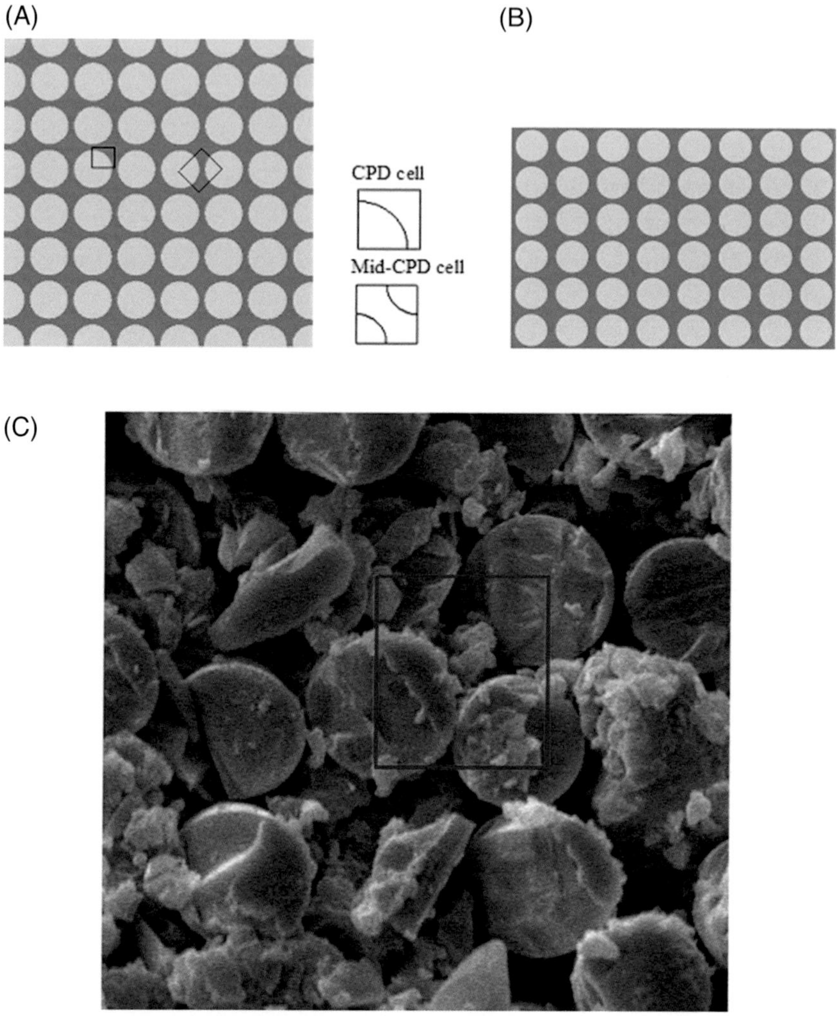

Figure 4.2 Types of fiber packing (A) square array, (B) rectangular array, and (C) scanning electron microscope photograph for square packing.

Rather, different hardening characteristics between the unidirectional composites with regular and random fiber packing have been found in the plastic regime. Karami and Garnich [8] investigated a finite element-based micromechanical model for fibrous materials and predicted the thermoelastic behavior of continuous fiber composites with fiber waviness. A periodic unit cell based on hexagonal fiber packing is assumed as a RVE. The orthotropic stiffness and thermal expansion parameters of the wavy fiber material are predicted by volume averaging stresses and strains. Gusev et al. [9] evaluated the effect of fiber packing and elastic properties of a transversely random unidirectional glass/epoxy composite.

Fiber packing is decided by the arrangement of fiber in a composite. It may be regular or irregular. Regular arrangement of fiber is classified as rectangular, square, and hexagonal packing. When fibers are arranged in irregular way, it may be known as novel packing.

(a) Traditional fiber packing

Assuming a periodic array of fibers as shown in Fig. 4.2, a typical repeating unit can be easily isolated by selecting the four closest symmetry axes, with a set of two parallel axes perpendicular to another set of two parallel axes. In Fig. 4.2, two different packing directions depending on the (transverse) loading direction can be defined. One is called the closest packing direction (CPD), the other the mid-closest packing direction (mid-CPD). The rectangular cell shows a strong anisotropy in the calculated global stress-strain response. A large difference in calculated strains is observed between the two loading directions.

These are periodic/uniform packing geometries and reduce the mechanical analysis of unidirectional fiber-reinforced composites to the solution of a set of boundary problems for repeating cells (or RVE). The surface is defined in the infinite medium by representing the composite material, which isolates a volume composed of one or several fibers.

Here, the volume (the unit cell or RVE) is assumed to be representative of the rest of the material, and the structure of the remaining part of the material is obtained by repeating the cell. Fig. 4.2 illustrates this assumption. The principle of repeating cells for periodic packing geometry shows symmetry of properties, which results in simple boundary conditions. A more uniform packing geometry is the hexagonal array packing, shown in Fig. 4.3A. This fiber packing geometry exhibits almost complete isotropy. Similar to the rectangular packing case, two packing directions, that is, CPD and mid-CPD, exist. Both directions have been used as loading directions. However, uniform cell models are fundamentally incorrect as, in actual practice, fibers are rarely packed in uniform fashion.

(b) Novel packing geometry

Novel fiber packing geometry proposed by Zhang et al. [6] shown in Fig. 4.4 is more realistic as it considers nonuniform distribution of fibers in matrix which causes severe nonuniform stress and strain distribution within the composite. The fiber rich area shows a relatively

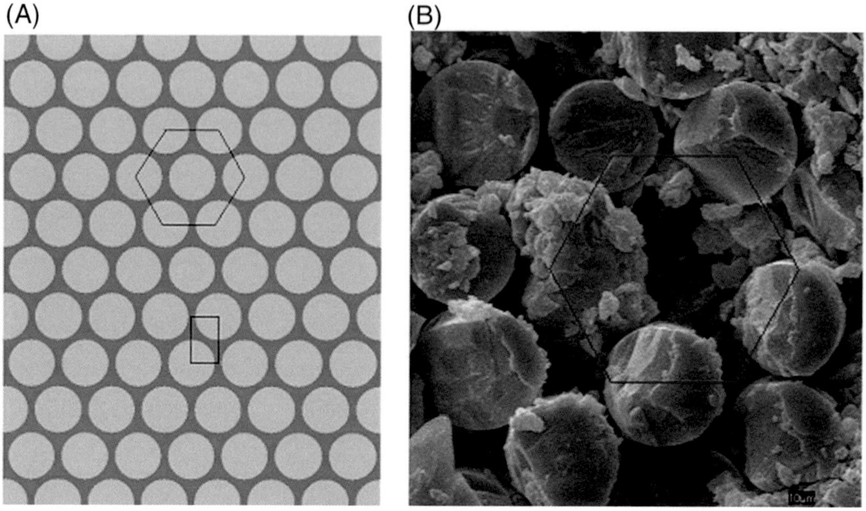

Figure 4.3 Types of fiber packing (A) hexagonal array and (B) scanning electron microscope photograph for hexagonal packing.

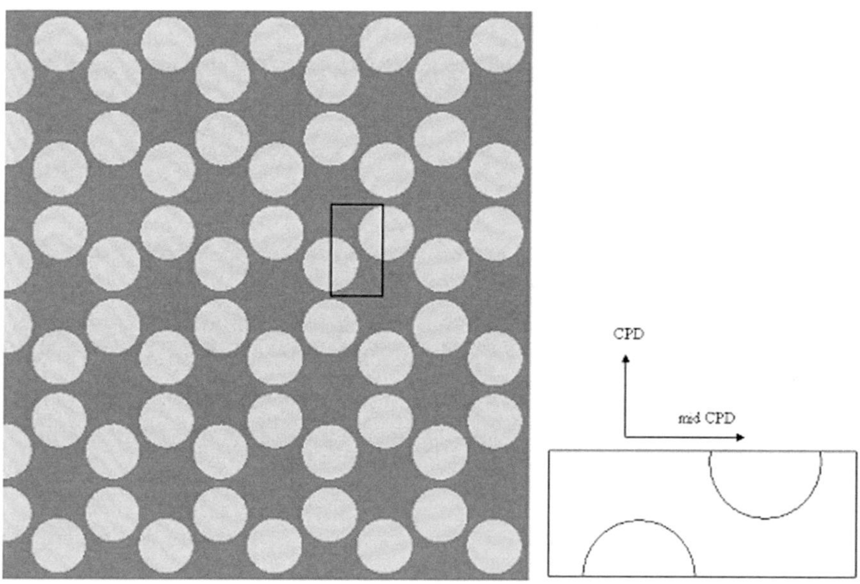

Figure 4.4 Novel packing array.

higher average stress level compared to matrix-rich area. In order to account for the influence of fiber-rich regions and matrix-rich regions, a modified regular fiber distribution pattern is proposed in the said work. Normally in filament wound fiber-reinforced composites, all the fibers are distributed close to each other forming clusters in which the distance between two neighboring fibers is usually smaller than the scale of fiber radius. The fiber rich area dominates the cross-section of the composite. Between the fiber clusters, some narrow-elongated matrix-rich regions are observed extending mostly parallel to the axis of the composite tube. The novel packing geometries based on above observations are globally transversely isotropic. Here the fibers and matrix are periodically arranged but with a local variation of fiber density. Therefore, these packing geometries are referred to as locally nonuniform. Novel packing geometry provides highest fiber volume fraction of 68%.

4.3 Finite element approach

Micromechanical damping analysis of unidirectional fiber-reinforced composites involves the determination of the contribution of three constituents that is, fiber, matrix, and fiber-matrix interface. Here, the state and properties of interfacial region are neglected and it is presumed that there is perfect bonding between fiber and matrix. The strain energy method [10] expresses, for a given loading, the composite loss factor as the ratio of summation over all elements of the structure of the product of the loss factor for each element and the strain energy for each element to the total strain energy.

The loss factor is expressed as,

$$\eta = \frac{\sum_{i=1}^{n} \eta_i W_i}{\sum_{i=1}^{n} W_i} \tag{4.41}$$

Micromechanical modeling by FEM takes into consideration the interaction of constituent materials. Thus, under particular loading, the strain energy is stored in both matrix and fiber. However, much of the energy dissipation comes from matrix deformation, though some energy dissipation occurs in fibers as well. Thus, for two-phase micromechanical model, the composite loss factor can be written as

$$\eta_c = \frac{\eta_f W_f + \eta_m W_m}{W_f + W_m} \tag{4.42}$$

The fiber packing in composite material for the analysis is considered to be of square array type. Thus, the composite material is assumed to be made up of RVE which repeats periodically. The strain energy for

Table 4.1 Material properties of constituents of two-phase GFRE composite.

Properties	E-glass fiber	Epoxy matrix
E	72.4 GPa	2.76 GPa
G	30.2 GPa	1.02 GPa
ν	0.2	0.35
η	0.0018	0.015

RVE is given as

$$W = \frac{1}{2}\int_V \sigma_{ij}\varepsilon_{ij}dV = \frac{1}{2}\sum \sigma_{ij}\varepsilon_{ij}\delta V = \sum \sigma_{ij}^T S_{ij}\sigma_{ij}\delta V \tag{4.43}$$

Thus, the strain energy for composite is written as

$$W_c = \sum (W_f + W_m) \tag{4.44}$$

where, strain energy stored in fiber and matrix is given by

$$W_f = \frac{1}{2}\sum \{\sigma_{ij}\}_f^T \{S_{ij}\}_f \{\sigma_{ij}\}_f \delta V \tag{4.45}$$

$$W_m = \frac{1}{2}\sum \{\sigma_{ij}\}_m^T \{S_{ij}\}_m \{\sigma_{ij}\}_m \delta V \tag{4.46}$$

Any particular "finite element" of the RVE is either made of fiber or matrix material. Accordingly, its strain energy is added to W_f or W_m. In order to implement FEM analysis to predict loss factors, the basic elastic and damping properties of the constituents of glass fiber-reinforced composite (GFRC) used are listed in Table 4.1.

4.3.1 FEM modeling with fiber packing geometry

Finite element method models considering RVE made up of SC square array (MCPD cell) and hexagonal packing are analyzed under different types of loading to study the effect of fiber packing on damping. Representative FEM models incorporating these packing geometries are shown in Fig. 4.5. Four-nodded quadrilateral elements are used to construct the 2D FEM models as shown in Fig. 4.5. Double fiber mid-CPD cell (square array shown in Fig. 4.5A) and hexagonal array in Fig. 4.5B are analyzed for fiber volume fractions $V_f = 0.4, 0.5, 0.6$, and 0.7. Four loss factors are determined using Eq. (4.42).

Finite element models for transverse normal, longitudinal normal, transverse shear, and longitudinal shear to predict damping in fiber-reinforced

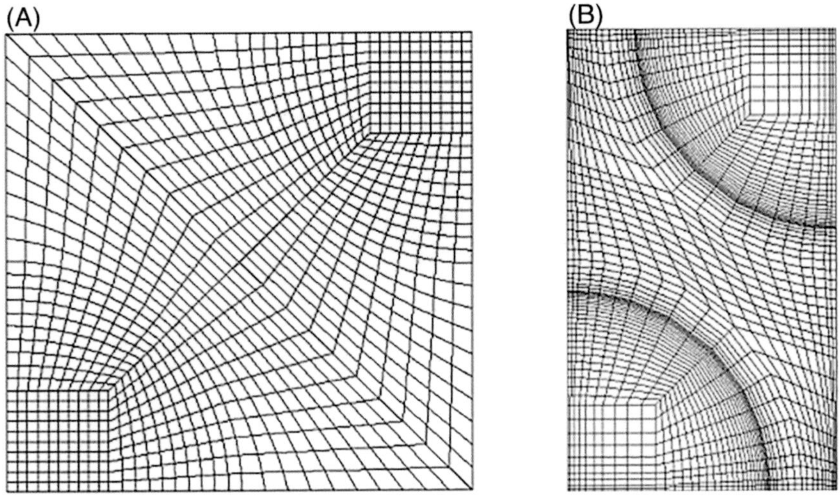

Figure 4.5 (A) Double fiber (mid CPD cell) model for square packing ($V_f = 0.4$) and (B) Hexagonal packing.

composites are developed in this section. Single cell (SC) and multi-cell (MC) RVE have been used.

D = Displacement, R = Rotation, F = free (no displacement and no constraint) for the prediction of elastic moduli of fiber-reinforced composites. FEM models for GFRC with various fiber volume fractions for fiber having cylindrical cross-section are constructed and subjected to different types of loading conditions. 3-D FEM models for GFRC with $V_f = 0.4, 0.5, 0.6,$ and 0.7 are constructed. Quarter domain 3-D FEM models with two fibers indicating the boundary conditions with longitudinal (11), transverse (22), longitudinal shear (12), transverse shear loading (23) is shown in Figs. 4.6 to 4.9. These FEM models incorporate hexahedron eight nodded finite elements. The boundary conditions of displacement type applied to the FEM models with longitudinal (11), transverse (22), longitudinal shear (12), and transverse shear (23) loadings are given in Table 4.2. An additional constraint has been applied to fix the origin. It may be noted that the axis z, y, and x of the FEM model refer to material axis 1, 2, and 3 respectively. In addition to the xy plane, the boundary conditions pertaining to yz and zx planes are available in Table 4.2. These boundary conditions are in reference to the plane of symmetry and the rear plane xy. For the shear loadings (12 and 23), (Figs. 4.8–4.9) all the six degrees, of freedom of nodal points lying in the xy-plane are restricted. The loading conditions for transverse shear (23)

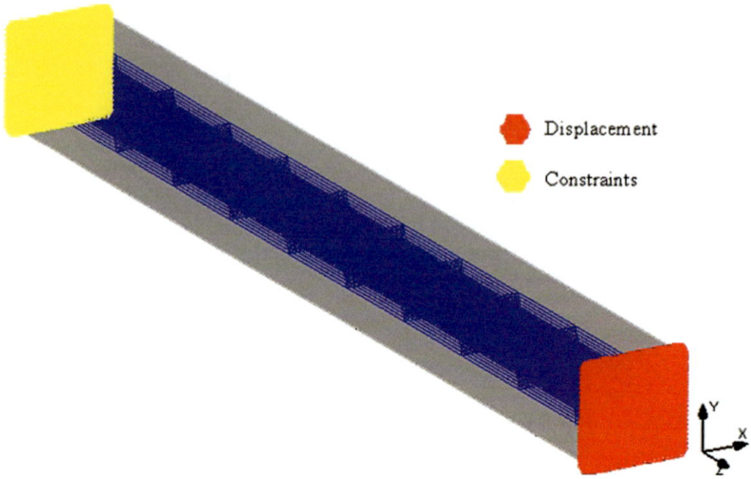

Figure 4.6 Boundary condition for longitudinal loading two fiber model mid-CPD with displacement boundary condition.

Table 4.2 Boundary condition for quarter model and hexagonal model of FEM modeling.

	Dx	Dy	Dz	Rx	Ry	Rz
longitudinal loading						
Constraints	F	F	0	0	0	F
Displacement			0.001			
Transverse loading						
Constraints yz plane	F	0	F	0	F	0
Constraintsxz plane	0	F	F	F	0	0
Displacement	0.001					
Transverse shear loading						
Constraints	0	0	0	0	0	0
Displacement	0.001	0	0	0	0	0
Longitudinal shear loading						
Constraints	0	0	0	0	0	0
Displacement	0	0	0.001	0	0	0

and longitudinal shear can also be well understood from Figs. 4.8 and 4.9. Accordingly, 3-D FEM models are constructed and analyzed using NISA 15 finite element software. These FEM models provide strain energy for the fiber and the matrix for the corresponding loading condition. Strain energy variation pattern in RVE is shown in Figs. 4.11–4.14 for longitudinal normal, transverse normal, transverse shear, and longitudinal shear loading conditions

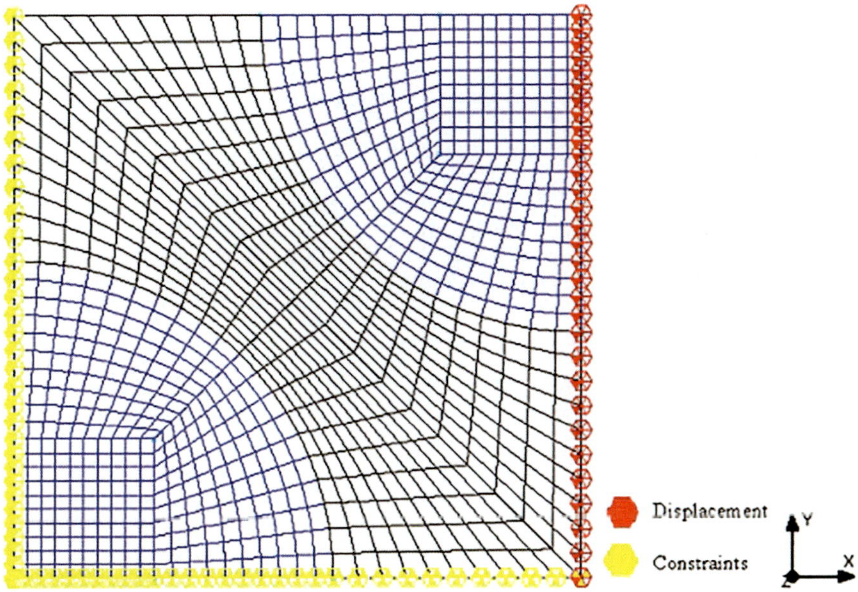

Figure 4.7 Boundary condition for transverse loading two fiber model mid CPD with displacement boundary condition.

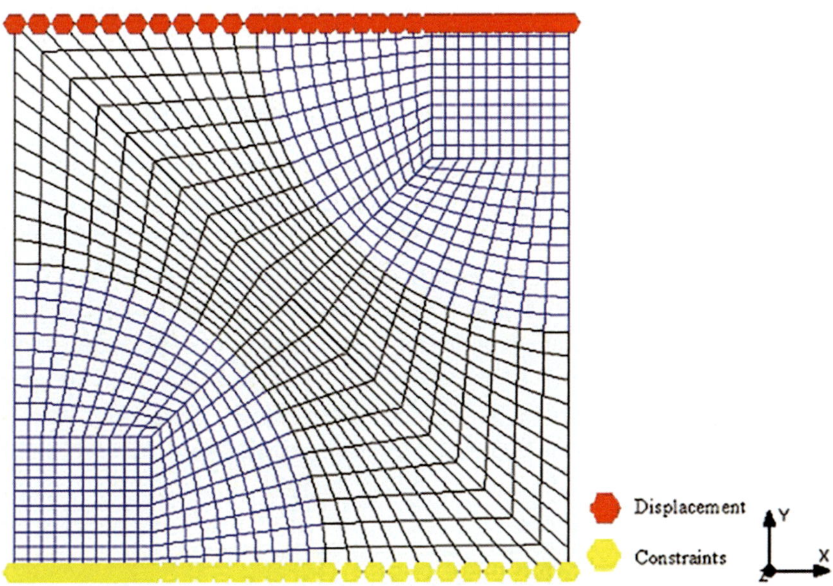

Figure 4.8 Boundary condition for transverse shear loading for two fiber model mid CPD with displacement boundary condition $V_f = 0.5$.

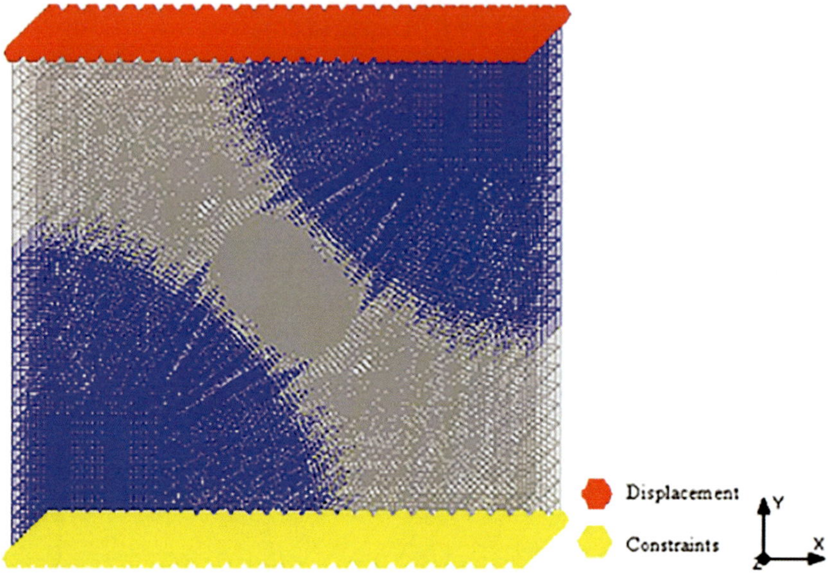

Figure 4.9 Boundary condition for longitudinal shear loading for two fiber model mid CPD with displacement boundary condition $V_f = 0.6$.

respectively. Four damping coefficients namely $\eta_{11}, \eta_{22}, \eta_{12}$ and η_{23} are thus determined using Eqs. (4.42), (4.45), and (4.46).

4.4 Mathematical model for frequency dependence

The proceeding section described the different types of models which have been implemented to evaluate the loss factors of composites. A comparative study of the results was conducted in detail. Another aspect of damping covered in this chapter relates to dependence of loss factors on frequency. In this section mathematical model for frequency dependence has been developed and the dependence of damping on frequency has been illustrated. The mechanical properties of these damping materials are often sensitive to frequency, temperature, type of deformation (i.e., shear or dilation), and sometimes amplitude. Material properties appropriate to each single temperature of interest are used in these analyses. Viscoelastic model is used for better accuracy, but it is not widely used yet. The model strain energy method proposed by (Roger et al., 1981) [20] is the most commonly used approach for the analysis of damping design of viscoelastic polymers. Lesieutre [11] modeled frequency dependent longitudinal dynamic behavior of linear viscoelastic long fiber composites. In this approach, the total

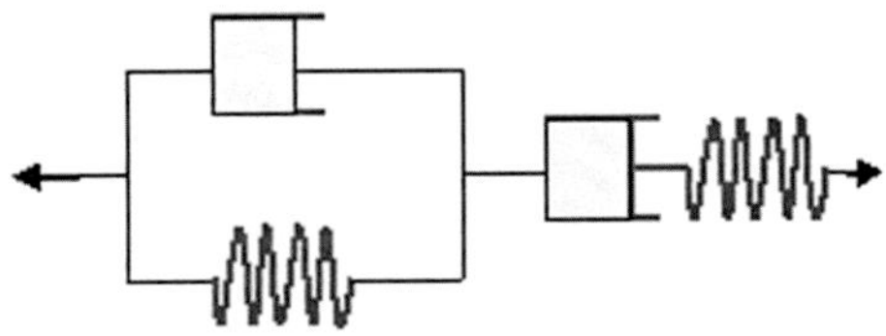

Figure 4.10 Four-parameter model for viscoelastic matrix.

displacement field is considered to be comprised of two parts: an elastic part and an anelastic part. Material dynamic behavior is described by constituent equations and a governing differential equation. Hadi and Ashton [12] evaluated material damping of a unidirectional composite experimentally by the flexural resonance method using cantilever beam specimens with different composite fiber orientations and for each fiber orientation at three fiber volume fractions over a frequency range of 100–1000 Hz. A finite element model is also used to predict the damping properties.

Chia et al. [13] measured the dynamic behavior of plasma sprayed NiCoCrAlY coatings using a dynamic mechanical analyzer (DMA) at low frequencies (0.01–0.1 Hz), indicating that the coatings exhibited viscoelastic behavior. Wei et al. [14] measured the damping behavior of foamed commercially pure aluminum. The internal friction and relative dynamic modulus have been measured at low frequency using multifunctional internal friction apparatus. Jaouen et al. [15] presents a simplified numerical model based on the hierarchical trigonometric functions set to predict the low frequency vibration behavior of a plate backed by a thin foam layer.

Maxwell-Voigt based four parameter model is applied to the bridging model which incorporates the effect of fiber packing, in order to study the effect of frequency.

4.4.1 Mathematical formulation

Matrix material is modeled as viscoelastic and the reinforcement fiber is assumed to be elastic. The behavior of viscoelastic materials is predicted by a model built from viscous elements (dashpot) and discrete elastic elements (spring). By combining the two springs and two dashpots, four parameter model [16] for viscoelastic material is formed as shown in Fig. 4.10. In this model, one dashpot and one spring are coupled in parallel, known as the Voigt model and joined in series with the Maxwell model (a serial coupling of linear spring and viscous damper). A differential equation is derived for the four-parameter viscoelastic model as given in Eqs. (4.47) is solved, and the elastic modulus of epoxy is estimated.

$$\sigma + p_1\dot{\sigma} + p_2\ddot{\sigma} = q_1\dot{\varepsilon} + q_2\ddot{\varepsilon} \tag{4.47}$$

where

$$p_1 = \frac{\mu_m}{E_m} + \frac{\mu_v}{E_v} + \frac{\mu_m}{E_v}, \; p_2 = \frac{\mu_m\mu_v}{E_m E_v}, \; q_1 = \mu_m, \; q_2 = \frac{\mu_m\mu_v}{E_v} \tag{4.48}$$

E_m, E_v are Maxwell and Voigt elastic modulus and μ_m, μ_v are Maxwell and Voigt viscosity of assumed four parameter model of viscoelastic material. Solving this Eq. (4.48) the storage modulus (E) and loss modulus (E′) have been obtained as given in Eqs. (4.49) and (4.50),

$$E = \frac{\omega^2\left(q_1 p_1 + \left(p_2\omega^2 - 1\right)q_2\right)}{1 + \omega^2 p_1 + \left(p_2\omega^2 - 2\right)p_2\omega^2} \tag{4.49}$$

$$E^* = \frac{\omega\left(q_1 + q_2 p_1\omega^2 - q_1 p_2\omega^2\right)}{1 + \omega^2 p_1 + \left(p_2\omega^2 - 2\right)p_2\omega^2} \tag{4.50}$$

Using the bridging model Eqs. (4.37)–(4.40) elastic moduli of the composite are evaluated. All the four loss factors $\eta_{11}, \eta_{22}, \eta_{12}, and\ \eta_{23}$ are predicted using the viscoelastic correspondence principal.

4.5 Results and discussions

Variation of elastic modulus and loss factors are predicted with increase of fiber volume fraction by varying the fiber packing factor using the materials properties given in Table 4.1. Optimization has been accomplished to obtain the equivalence FEM fiber packing factor for a particular volume fraction. Further dependence of loss factor on frequency has been obtained.

4.5.1 Prediction of strain energy

Figs. 4.11–4.14 show the variation of strain energy for the RVE of fiber reinforced composites ($V_f = 0.4$) for all four types of loading, namely longitudinal, transverse, transverse shear, and longitudinal shear. The variation of strain energy for longitudinal loading is shown in Fig. 4.11. Observation indicates that the variation of strain energy is mostly in fiber, and strain energy variation remains constant in the matrix. Variation of strain energy is more in matrix for transverse normal loading as shown in Fig. 4.12. Strain energy is very low in fiber because strain is very less in fiber as its stiffness is very high in comparison to the matrix. Fig. 4.13 depicts the variation of strain energy is maximum at the interface of fiber and matrix for transverse shear loading. The variation of strain energy is shown in

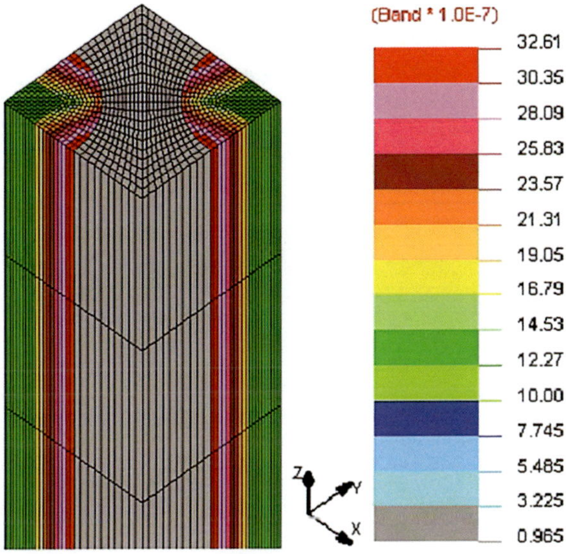

Figure 4.11 Strain energy for longitudinal normal loading.

Fig. 4.14 for longitudinal shear loading. Variation of strain energy is maximum in matrix and almost constant in fiber.

The loss factor of a composite is obtained by the Eq. (4.41) using the material properties given in Table 4.1. These loss factors obtained are compared with the bridging model results and other published work.

4.5.2 Effect of fiber volume fraction

Predicted results based on the bridging model for elastic modulus and damping are presented here. The loss factors predicted by the FEM/strain energy approach is compared with the bridging model results. Finally, a comparative study is conducted to correlate $\eta_{11}, \eta_{22}, \eta_{12}$ and η_{23} evaluated by various methods. Elastic moduli and loss factors are predicted for various value of the fiber volume fraction. The effect of fiber packing factors have been evaluated on the elastic moduli and loss factors.

The various elastic moduli and loss factors have been predicted by different methods described in Section 4.1. Computer code in MATLAB has been written to predict the elastic moduli based on a two-phase bridging model and the results so obtained are compared with published work [3] as given in Table 4.3.

(a) Elastic moduli E_{11} and E_{22}

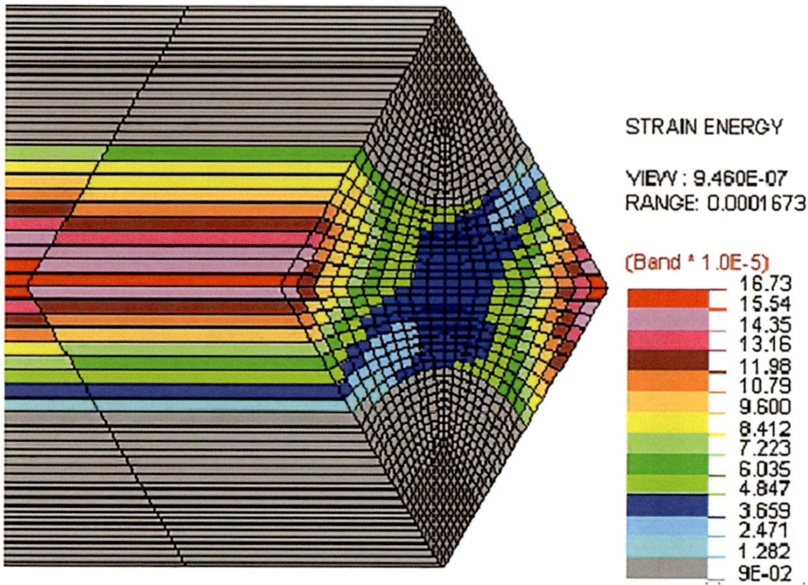

Figure 4.12 Strain energy for transverse normal loading.

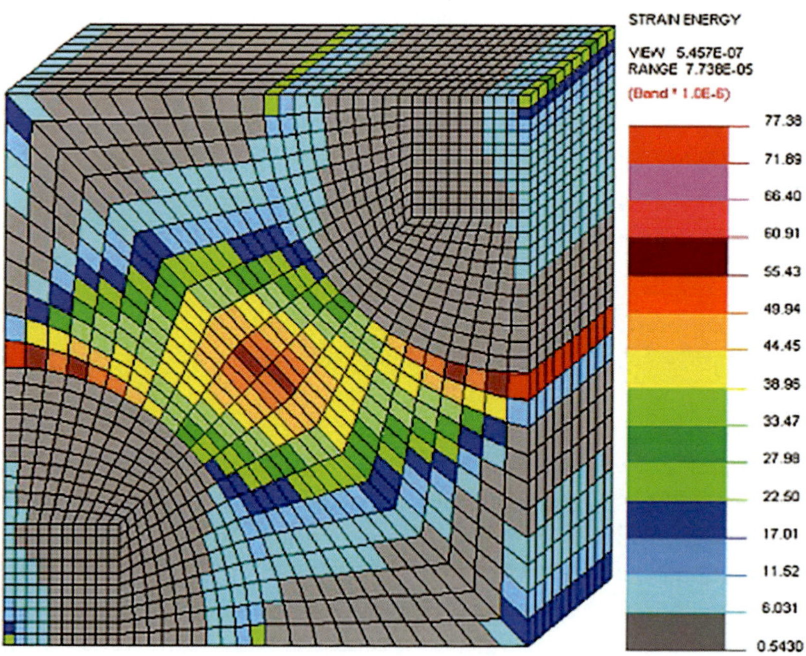

Figure 4.13 Strain energy for transverse shear loading.

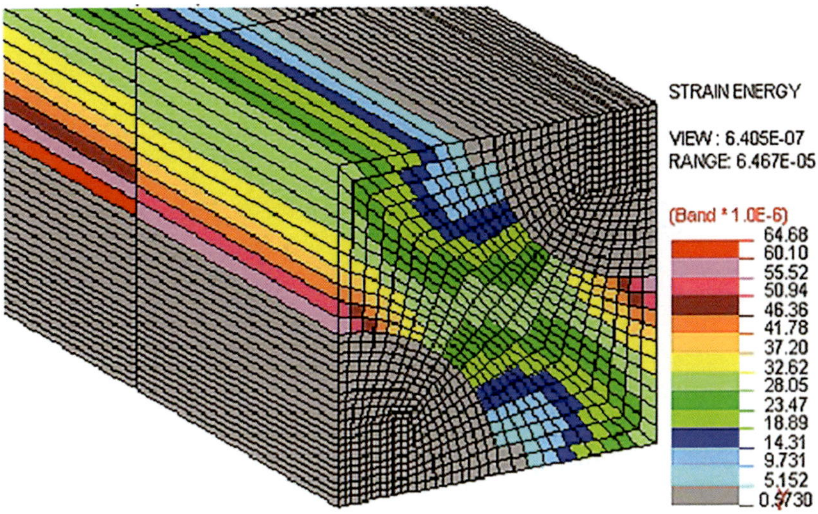

Figure 4.14 Strain energy for longitudinal shear loading.

Table 4.3 Elastic modulus of fiber-reinforced composite.

Properties	E Glass 21 x K43 Gevetex (GPa)	Matrix (GPa)	Composite $V_f = 0.62$ [3]	Composite $V_f = 0.62$ (Author results)	Composite $V_f = 0.62$ (Soden, 1998)
E_{11}	80	3.35	50.87	50.87	53.48
E_{22}	80	3.35	14.38	14.38	17.7
G_{12}	33.33	1.24	5.72	5.72	5.83
G_{23}	33.33	1.24	–	4.88	–
ν	0.35	0.35	0.257	0.257	0.278

Fig. 4.15 demonstrates the variation of the longitudinal of the elastic modulus E_{11} of composite with respect to fiber volume fraction (V_f). Variation of elastic modulus is predicted against the increase of longitudinal fiber packing factor (α) and transverse packing factor (β). Longitudinal elastic modulus (E_{11}) shows insensitiveness to the fiber packing factors α and β as fiber packing factors vary from 0.1 to 1.0. In this case, fiber and matrix are stressed in longitudinal direction and not much interaction takes place between fiber and matrix. Therefore, arrangement of fiber and matrix does not have any role in taking the load. The longitudinal elastic modulus (E_{11}) as predicted by the models bridging model, Eshelby's method, and Servanos-Chamis approach is practically the same and varies linearly with fiber volume fraction.

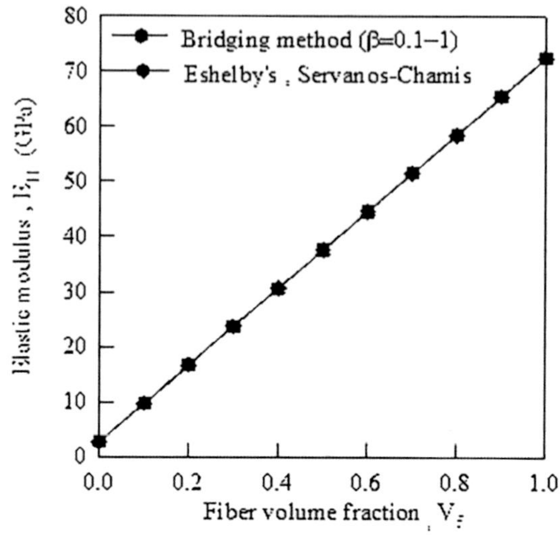

Figure 4.15 Variation of longitudinal elastic modulus E_{11} of composite with respect to fiber volume fraction (V_f).

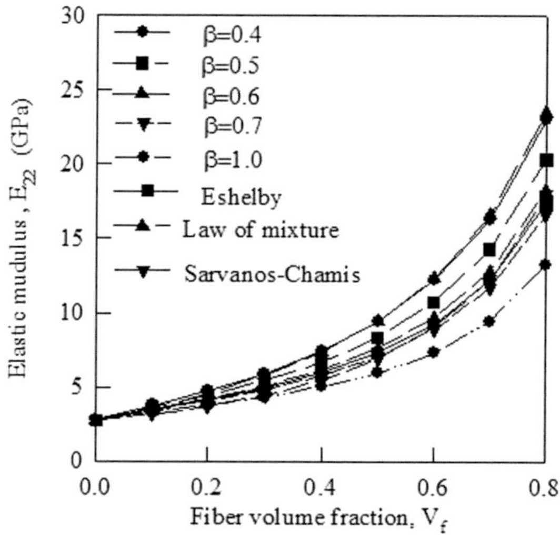

Figure 4.16 Variation of longitudinal elastic modulus E_{22} of composite with respect to fiber volume fraction (V_f).

Fig. 4.16 depicts the comparison of the results for transverse elastic modulus E_{22} with respect to fiber volume fraction predicted by the Eshelby method, Law of mixture, Servanos-Chamis approach, and the bridging model. The law of mixture results are the same as the bridging model results for value of $\beta = 0.4$. The results obtained for $\beta = 0.6$

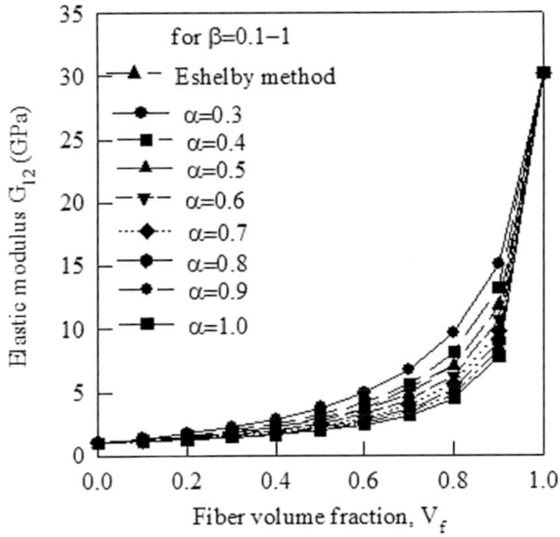

Figure 4.17 Variation of elastic modulus G_{12} of composite with respect to fiber volume fraction (V_f).

are almost equal to the results predicted by Eshelby and Servanos-Chamis for the range of fiber volume fraction V_f from 0.2 to 0.6. The transverse fiber packing factor assesses the closeness of the fibers in the composite and thus, for higher fiber volume fraction, there is corresponding increase in E_{22}.

(b) *Shear moduli G_{12} and G_{23}*

Fig. 4.17 shows the effect of fiber volume fraction V_f on shear modulus G_{12} and that it increases with the increase of fiber volume fraction. Various curves represent variation of G_{12} for longitudinal fiber packing factor $0.3 < \alpha < 1.0$. With the increase in the value of α, G_{12} decreases rapidly for lower range of α. Increase in longitudinal shear modulus is more for lower value of α as the fiber volume fraction increase. Loss in longitudinal shear modulus with increase in α is regular, that is, similar trend is obtained for all value of α. For longitudinal fiber packing factor α is equal to 0.5, the results are in good agreement with Eshelby results.

Fig. 4.18 demonstrates the variation of transverse shear modulus, G_{23} with respect to fiber volume fraction. Effect of β is very clear for higher volume fraction decrease in magnitude of G_{23} is more for lower value of β. It can be seen that for $V_f = 0.3$ decrease in magnitude of G_{23} is 17.71%, whereas for $V_f = 0.5$ increase in G_{23} is 28.63% for the corresponding variation of β from 1.0 to 0.6. Prediction of G_{23}

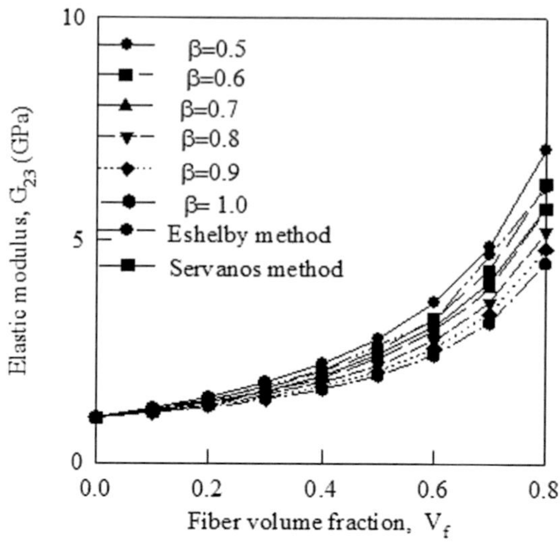

Figure 4.18 Variation of elastic modulus G_{23} of composite with respect to fiber volume fraction (V_f).

by the bridging model for $\beta = 0.5$ in the range of volume fraction $V_f = 0.2$–0.7 matches very well with the one obtained by Eshelby's method.

(c) *Prediction of loss factors η_{11} and η_{22}*
The results of loss factors using the bridging model are presented in this section and these results are compared with results predicted by FEM, Eshelby, and Sarvanos methods.

Figs. 4.19 illustrates the effect of V_f on the longitudinal normal loss factor (η_{11}) along with the fiber packing factor. The decrease in magnitude of loss factors is rapid for lower values of Vf and it shows insensitiveness towards the fiber packing factors α and β. In Fig. 4.19 loss factor (η_{11}) decreases with the increase of fiber volume fraction. The Bridging model predicted results for fiber packing factors α and β from 0.1 to 1.0 were almost similar to the results of Sarvanos, Hashin, and FEM.

Fig. 4.20 shows the predicted variation of the loss factor η_{22} with respect to fiber volume fraction using bridging model for different transverse fiber packing factorβ. The transverse loss factor decreases with an increase of transverse fiber packing factorβ. Observation indicates that for lower volume fraction ($0.1 < V_f < 0.4$) transverse loss factor increases as β increases from 0.3 to 1 in increment of 0.1. This increase in loss factor η_{22} is small increasing value of β corresponding to appropriate value of $V_f$between 0.1 and 0.8.

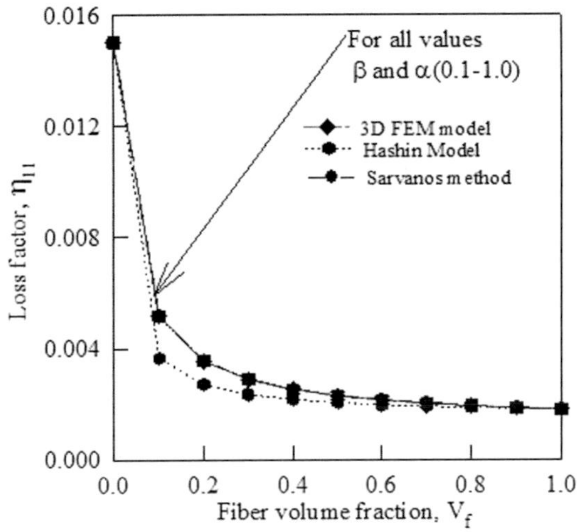

Figure 4.19 Variation of longitudinal loss factor η_{11}.

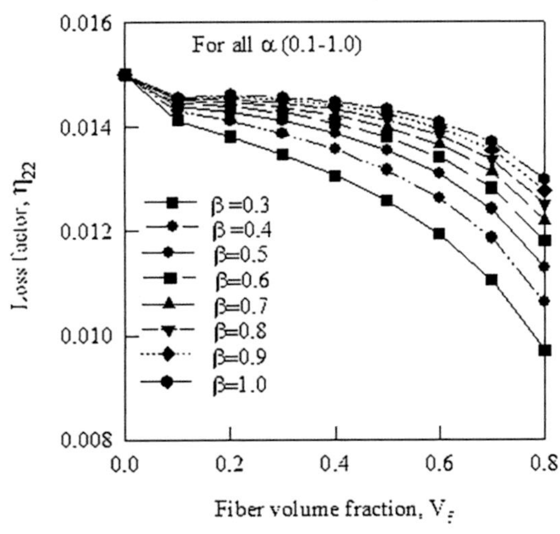

Figure 4.20 Variation of transverse loss factor η_{22}.

However, transverse loss factor η_{22} is very sensitive to transverse packing factor β for higher fiber volume fraction ($0.4 < V_f < 0.8$).

Fig. 4.21 shows the variation of the longitudinal shear loss factor, η_{12} with volume fraction. There is a steady decrease in η_{12} with increase in fiber volume fraction for higher fiber packing factorα. Fiber packing factor α plays a very significant role for the prediction of loss factor. Longitudinal

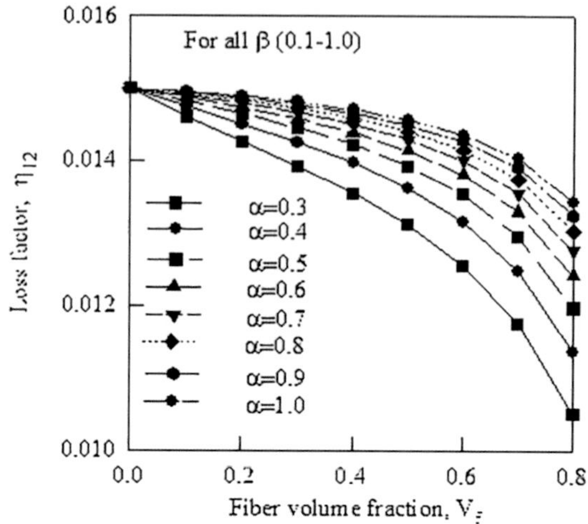

Figure 4.21 Variation of longitudinal shear loss factor η_{12}.

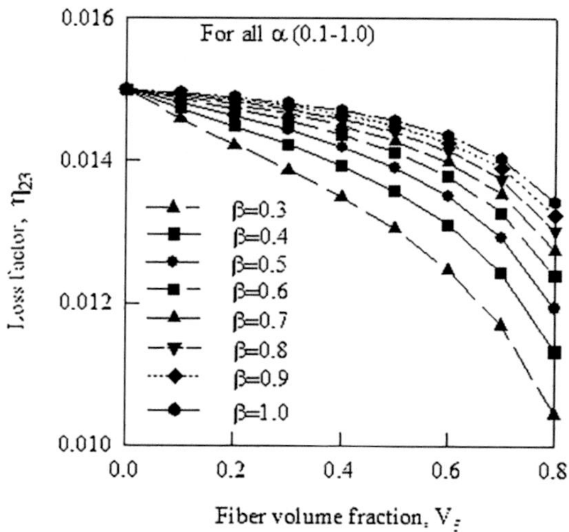

Figure 4.22 Variation of transverse shear loss factor η_{23}.

shear loss factor decreases at faster pace as longitudinal fiber packing factor α decreases from 1 to 0.3 at higher volume fraction between 0.4 and 0.8.

Variation of transverse shear loss factor η_{23} with respect to fiber volume fraction has been illustrated in Fig. 4.22 for different transverse fiber packing factorβ. Loss factor η_{23} decreases with increase of fiber volume fraction for particular fiber packing factor and this decrease is more for lower fiber packing factor (i.e., up to $\beta = 0.4$).

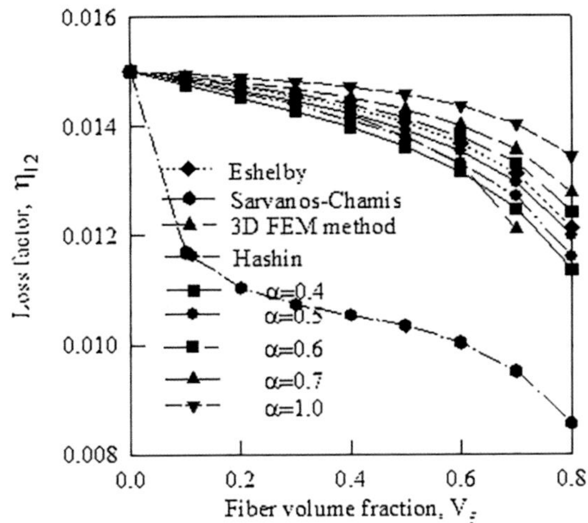

Figure 4.23 Comparison of longitudinal shear loss factor η_{12}.

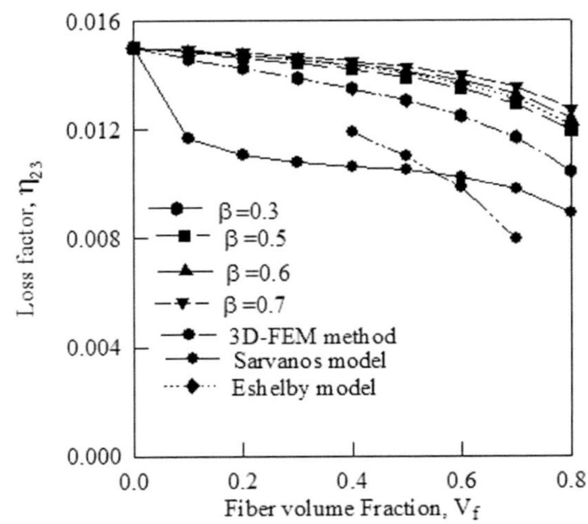

Figure 4.24 Comparison of transverse shear loss factor η_{23}.

Fig. 4.23 depicts the comparison of the longitudinal shear loss factor obtained from the bridging model with the result of Eshelby, Hashin, Sarvanos-Chamis, and FEM model. For $\alpha = 0.4$ results of bridging model are almost same as that of FEM model for fiber volume fraction from 40% to 60%.

Finite element method results for transverse shear loss factor are shown in Fig. 4.24. The bridging model results are in good agreement with Eshelby's model results. FEM results are comparable with the bridging model result for

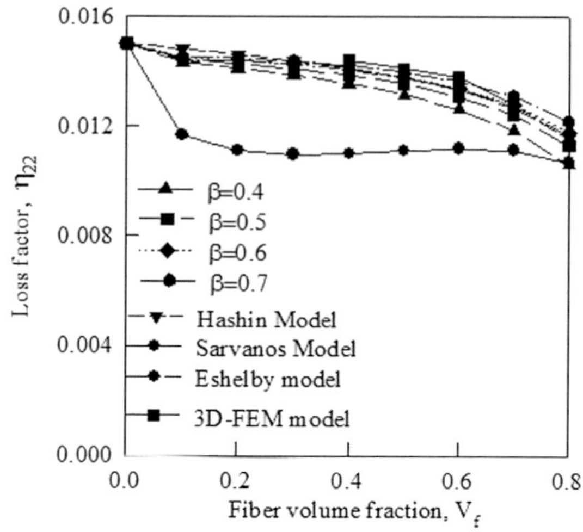

Figure 4.25 Comparison of transverse loss factor η_{22}.

lower value of $\alpha = 0.1$. Comparison of the results obtained by the bridging model with Eshelby, Hashin, Sarvanos-Chamis for transverse loss factor η_{22} are shown in Fig. 4.25. Results obtained by the bridging model for $\beta = 0.6$ are same as the Eshelby model results for the complete practical range of fiber volume fraction. The FEM results are slightly higher than the bridging model results.

4.5.3 Optimization of fiber packing factor α and β

Based on the variation of loss factors with fiber packing factor, a natural question arises as to what are the best values of packing factors that give accurate predictions one can use in the bridging model. The objective of the optimization of fiber packing factors α and β is to correlate the prediction of loss factors results as obtained from the bridging model with that of FEM models for different fiber packing factor. FEM model has been constructed for square and hexagonal packing and analyzed under different loadings for the prediction of corresponding loss factors. Further these results are compared with the bridging model results. Using the MATLAB optimization programme *fgoalattain,* the fiber packing factor for the particular packing is obtained. *Fgoalattain* is a multiobjective optimization programme as discussed below:

Function

[x, FVAL, ATTAINFACTOR, EXITFLAG, OUTPUT, LAMBDA]= fgoalattain (FUN, x, GOAL, WEIGHT, LB, UB, options)

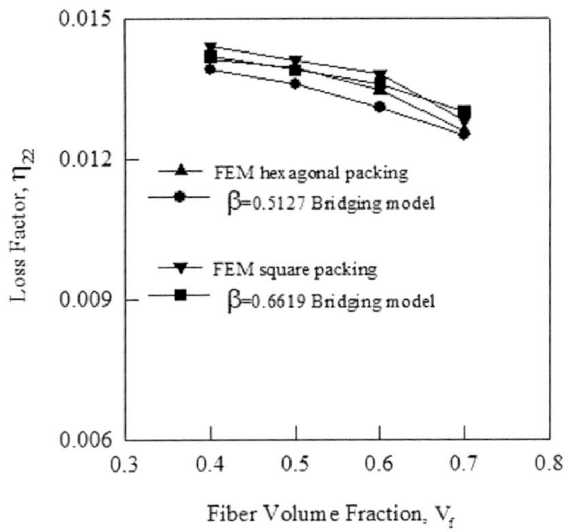

Figure 4.26 Variation of transverse loss factor with respect to V_f.

X = FGOALATTAIN (FUN, X0, GOAL, WEIGHT) tries to make the objective functions (F) supplied by the function FUN attain the goals (GOAL) by varying X. The goals are weighted according to WEIGHT. In doing so, the following nonlinear programming problem is solved:

$$\min\{\text{LAMBDA} : \text{F(X)}-\text{WEIGHT}.*\text{LAMBDA} <= \text{GOAL}\}$$

$$\text{X, LAMBDA}$$

$$\text{lb} <= \text{x} < \text{ub}$$

where x, weight, goal, lb, and ub are vectors, and F(x) are functions that return vectors. F(x), can be nonlinear. In this program X is fiber packing factor (β) and goal are loss factors. FUN consists of equation for calculating loss factor for different fiber volume fractions. Value of X initialized and loss factors are calculated, than compare with given value of loss factor. Range of x is 0.1 to 1.

Fiber packing factor characterization for loss factors

Loss factors have been evaluated and variation of the loss factor with respect to fiber volume fraction V_f for different value of fiber packing factor has been depicted in Figs. 4.26–4.28. Fig. 4.26 show that the loss factor (η_{22}) decreases with the increase of fiber volume fraction V_f for both hexagonal and square packing geometry. Variation of the bridging model results variation is similar to that of FEM results. FEM results for square packing geometry decrease drastically for higher value of V_f. For $\beta = 0.5127$

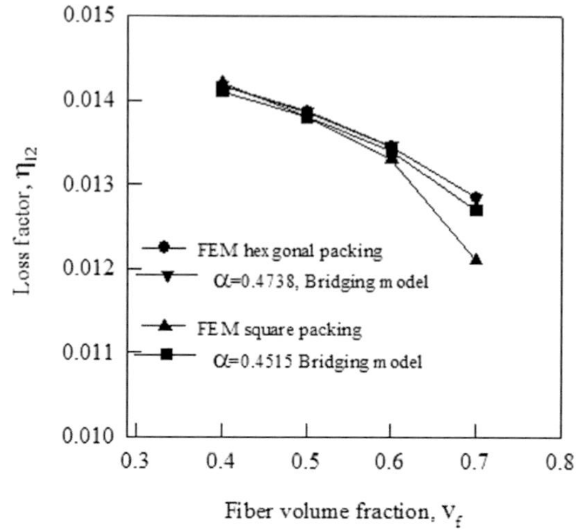

Figure 4.27 Variation of longitudinal shear loss factor with respect to V_f.

bridging model results are equivalent to hexagonal packing and when β is equal to 0.6619 for square packing.

Fig. 4.27 illustrates the comparison of the loss factor, η_{12} results in relation to fiber volume fraction V_f, predicted by the bridging model and FEM models. Longitudinal loss factor predicted by the bridging model for fiber packing factor $\alpha = 0.4738$ is same as that obtained from FEM model with hexagonal fiber packing. For square packing variation of loss factor is similar to FEM results up to 60% of fiber volume fraction at $\alpha = 0.4515$. It is concluded that if $\alpha = 0.4738$ fiber packing is hexagonal and square packing for $\alpha = 0.377$ while obtaining longitudinal shear loss factor.

Fig. 4.28 elaborates the variation of the loss factor, η_{23} with respect to fiber volume fraction V_f. It also provides insight into the effect of fiber packing factor β on loss factor, η_{23}. Observations indicate that loss factor, η_{23} is sensitive to fiber packing factor β indicating increasing trend as β increases from 0.133 to 0.141 for all fiber volume fractions, V_f. Further the loss factor, η_{23} results of FEM model with hexagonal packing geometry compare well with those obtained by bridging model for $\beta = 0.133$. It is appropriate to mention here, that in above cases, there exists a nonuniform state of stress due to the mismatch of properties of fiber, matrix, and fiber packing geometry. In the bridging model, the effect of fiber packing geometry under transverse and transverse shear loading is accounted for transverse packing factor β. Thus, the equivalence of fiber packing factor β effect in the bridging model

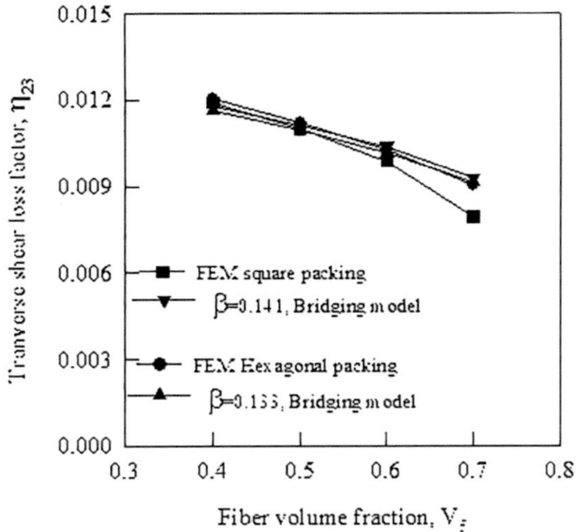

Figure 4.28 Variation of transverse shear loss factor with respect to V_f.

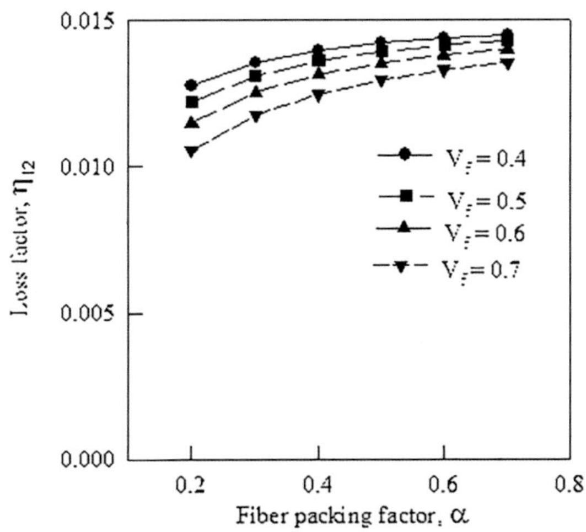

Figure 4.29 Variation of longitudinal shear loss factor with respect to packing factor.

to that of fiber packing geometry, that is, hexagonal packing geometry in FEM model is quite relevant. Prediction of loss factor is very similar to FEM results for square packing up to the $V_f = 0.6$, and here the bridging model results are almost equal to the FEM results.

Figs. 4.29–4.30 show the variation of loss factor with respect to fiber packing factor α and β for fiber volume fraction in the range of 0.4–0.7.

It can be seen from Fig. 4.29 that the loss factor, η_{12} increases with an increase in the magnitude of the longitudinal fiber packing factor α.

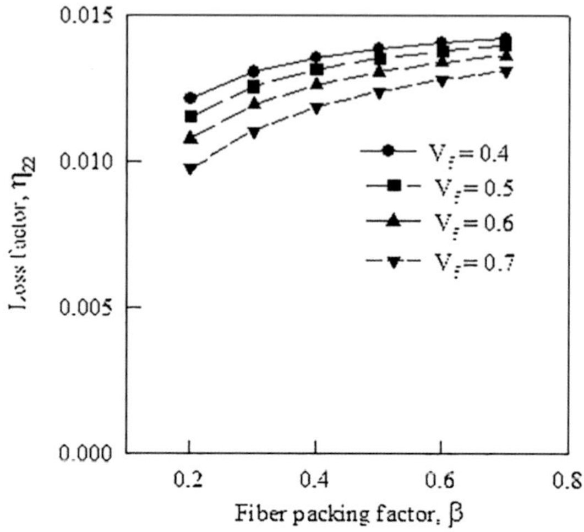

Figure 4.30 Variation of transverse loss factor with respect to packing factor.

Table 4.4 Characterization of fiber packing factor.

Loss factor	Packing factor	Square packing	Hexagonal packing
η_{22}	β	0.6619	0.5127
η_{23}	β	0.1410	0.1330
η_{12}	α	0.4515	0.4738

Fig. 4.30 provides the variation of loss factor η_{22} in reference to transverse fiber packing factor β for fiber volume fraction $V_f = 0.4 - 0.7$. Results for loss factors, η_{22} based on the bridging model show similar behavior as in the case of longitudinal shear loss factor, η_{12}. Fiber packing factors play significant role at higher fiber volume fractions because more fibers are available for proper distribution of fiber. The characterization of the fiber packing is given in Table 4.4.

4.5.4 Estimation of loss factors with frequency

The bridging model has been modified using the four-parameter viscoelastic model to consider the effect of frequency dependence of GFRE. Using the properties mentioned in Table 4.5 of glass fiber and epoxy resin, loss factors variation with respect to frequency has been evaluated. The parameters of the four-parameter model are same as those determined for epoxy by experimental results. The effects of fiber packing factor and fiber volume fraction have been evaluated.

Figs. 4.31–4.34 depict the variation of the loss factor with frequency from 1 to 100 Hz by varying the fiber volume fraction. Fig. 4.31 shows the variation

Table 4.5 Properties of the material used.

Matrix	$E_m = 4.56$ GPa $\mu_m = 759$ GPa.h	$E_v = 222.88$ GPa, $\mu_v = 178$ GPa.h	$\nu_0 = 0.35$
Fiber	$E_f = 72.6$ GPa	$\eta_f = 0.0018$	$\nu = 0.2$

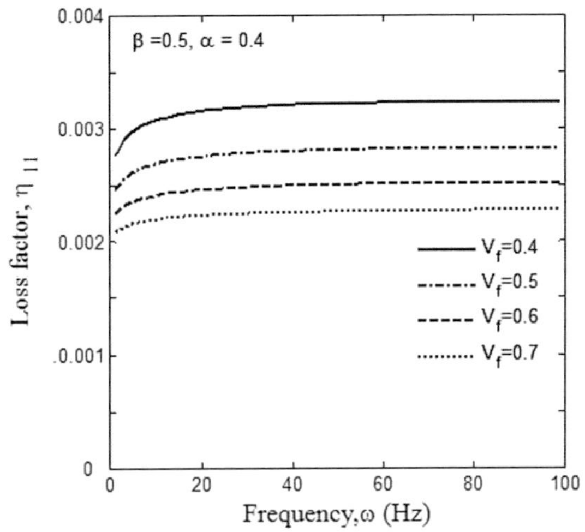

Figure 4.31 Variation of longitudinal loss factor with frequency for different fiber volume fraction.

of the longitudinal loss factor with frequency for fiber volume fraction from 0.4 to 0.7. The longitudinal loss factor increases first at very low frequencies (up to 5 Hz) and remains almost constant at higher frequency. Increase in loss factor is more for low fiber volume fraction in comparison to that higher. Loss factor η_{11} is independent of transverse fiber packing factor and longitudinal fiber packing factor as most of the load in longitudinal direction is taken by fiber only. Fig. 4.32 demonstrates the variation of the transverse loss factor with frequency for different fiber volume fraction. Decrease in loss factor, η_{22} is very sharp for lower values of frequency (up to 15 Hz) but for higher values of frequency, the reduction in loss factor is less. The decrease in transverse loss factor is more for lower V_f (i.e., for $V_f = 0.4$ decrease in η_{22} is 0.49%) than for the higher fiber volume fraction (i.e., for $V_f = 0.7$ decrease in η_{22} is 0.46%). Fig. 4.33 shows the variation of longitudinal shear loss factor η_{12} with frequency and it decreases with increase of frequency. A similar trend of variation is also observed for transverse shear loss factor as depicted in Fig. 4.34. Decrease in shear loss factor is due to dominating effect of matrix over the fiber in shear but fiber is playing the important role

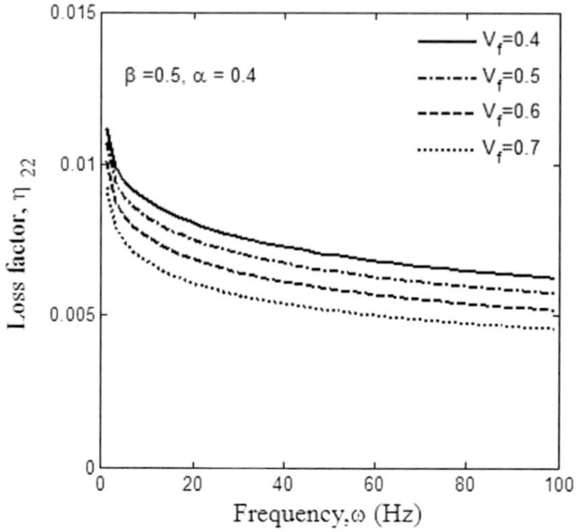

Figure 4.32 Variation of longitudinal shear loss factor with frequency for different fiber volume fraction.

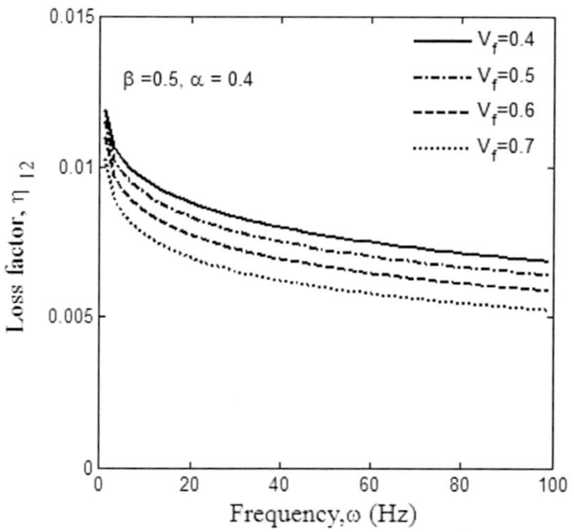

Figure 4.33 Variation of transverse loss factor with frequency for different fiber volume fraction.

in longitudinal loss factor and hence longitudinal loss factor increase with increases of frequency.

Figs. 4.35–4.37 demonstrate the variation of loss factors with increase of frequency incorporating the effect of fiber packing factor α and β for $V_f = 0.4$. Transverse loss factor, η_{22} variations is shown in Fig. 4.35 and it decreases with the increase of frequency. Effect of transverse fiber packing

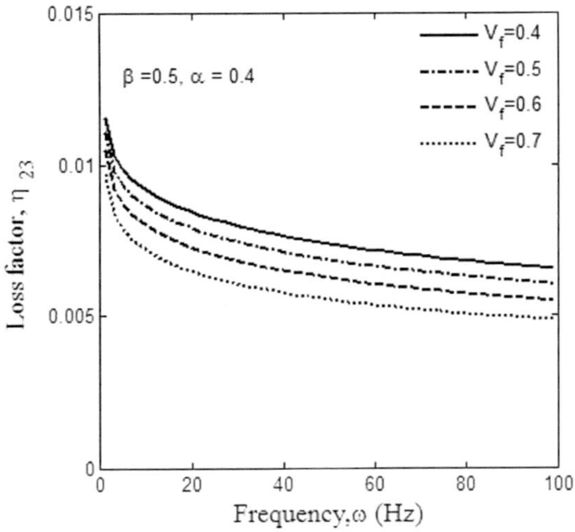

Figure 4.34 Variation of transverse shear loss factor with frequency for different fiber volume traction.

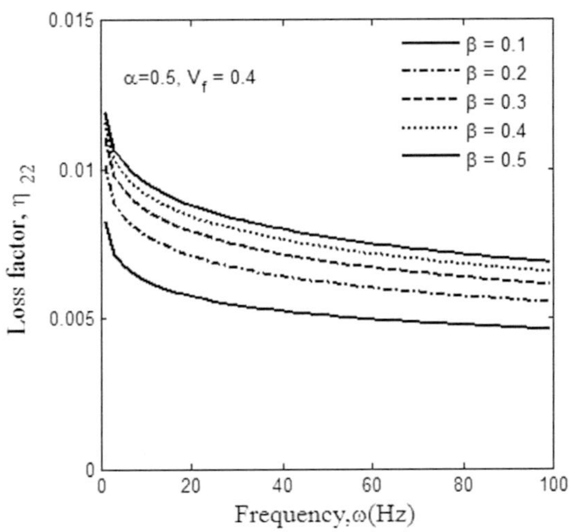

Figure 4.35 Variation of transverse loss factor with frequency for different fiber packing factor.

factor (β) is more at lower frequency. As the frequency increases variation due to β is keep on decreasing. Fig. 4.36 shows the variation of transverse shear loss factor with respect to frequency for different transverse fiber

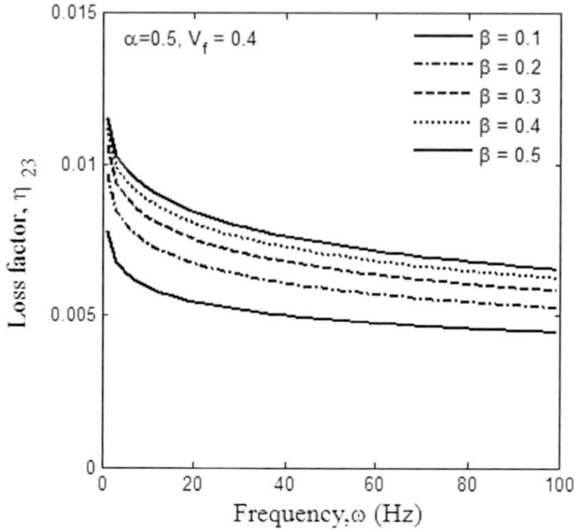

Figure 4.36 Variation of transverse shear loss factor with frequency for different fiber packing factor.

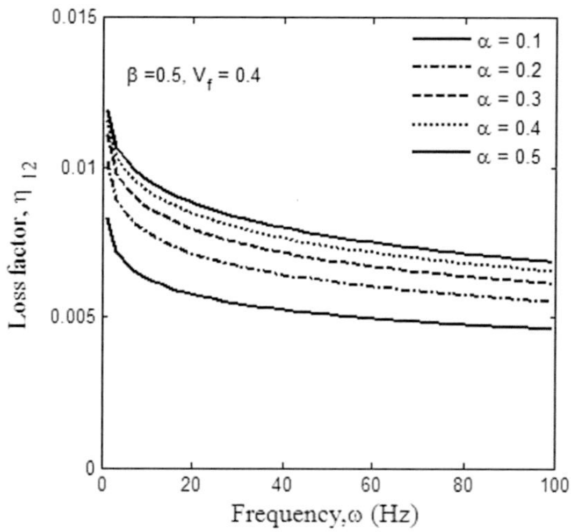

Figure 4.37 Variation of longitudinal shear loss factor with frequency for different fiber packing factor.

packing factor, and it decreases with the increase of frequency. Fig. 4.37 elaborates the effect of frequency on longitudinal shear loss factor and it decreases with increase in frequency. Loss factor, η_{12} magnitude increases with increase of longitudinal fiber packing factor (α) and its effect is more at lower frequency.

4.6 Conclusion

Elastic moduli and loss factors are predicted using the bridging model which includes the effect of fiber packing through the fiber packing factors. The results of the bridging model are compared with the Eshelby's method, Servanos–Chamis approach, FEM strain energy, and Hashin model. Optimization of fiber packing factor has been carried out for loss factors to correlate the result with those of FEM approach. The purpose of such a study is to identify the fiber packing geometry in relation to fiber packing factor α and β employed in the bridging model. Four parameter viscoelastic model is applied to the bridging model to incorporate the effect of frequency in addition to fiber packing effect. Based on the studies carried out the following conclusions are drawn:

(i) *Elastic modulus*

The following conclusions are made based on the comparative study for elastic moduli for varying the fiber volume fraction

a) Longitudinal elastic modulus, E_{11} increases linearly with the increase of fiber volume fraction. Longitudinal (α) and transverse (β) fiber packing factor has no effect on the longitudinal elastic modulus.

b) The transverse normal modulus, E_{22} increases with the improvement in fiber volume fraction. It increases very little for a lower value of V_f but for a higher value of V_f, it increases very rapidly. E_{22} increases with an increase in transverse fiber packing factor (β).

c) Longitudinal shear modulus is sensitive to longitudinal fiber packing factor (α) for lower fiber volume fraction ($V_f =$ up to 0.4) whereas more sensitive for higher fiber volume fractions ($0.4 < V_f < 0.9$) as fiber packing factor α decreases from 1 to 0.3.

(ii) *Loss factors*

Loss factors ($\eta_{11}, \eta_{22}, \eta_{12},$ and η_{23}) predicted by the bridging model are compared with other models and the following conclusions are made:

a) Longitudinal loss factor, η_{11} indicates decreasing trend with the increase of fiber volume fraction. Longitudinal loss factor is independent of both longitudinal and transverse fiber packing factor. There is one to one correspondence of results for longitudinal loss factor as predicted by FEM.

b) Transverse, transverse shear, and longitudinal shear loss factors decrease with the increase of fiber volume fraction.

c) Transverse loss factor (η_{22}) and transverse shear loss factor (η_{23}) increases with increase of fiber packing factor β corresponding to particular fiber volume fraction.

d) Longitudinal shear loss factor increases with the increase of longitudinal fiber packing factor (α) corresponding to particular fiber volume fraction.

e) Transverse loss factor, transverse shear loss factor, longitudinal shear loss factor has been obtained using FEM methods for hexagonal and square fiber packing. Using the bridging model, loss factors results are obtained and compared with FEM results and fiber packing factors α and β has been obtained with the help of optimization.

f) Optimal values of β are 0.6619 for transverse loss factor and 0.141 for transverse shear loss factor and $\alpha = 0.4515$ for longitudinal shear loss factor with the recommended values the results of the bridging model are closest to FEM results.

(iii) *Frequency dependence*

a) The bridging model and four parameter viscoelastic model for matrix has been combined and frequency dependence of loss factors are evaluated.

b) The longitudinal loss factor (η_{11}) increases with the increase of frequency rapidly and then almost remains constant for higher values of frequency.

c) The loss factors ($\eta_{22}, \eta_{12,}$ and η_{23}) decrease rapidly with an increase of frequency, however its reduction is very less at higher values of frequency. With an increase of fiber packing factor loss factors, increase and play a very significant role. Further, loss factors show an increasing trend with increase in fiber packing factors. A decrease in loss factors has been observed with an increase in volume fraction and frequency.

References

[1] Z. Hashin, Complex moduli of viscoelastic composites-I. General theory and application to particulate composites, Int. J. Solids Struct. 6 (5) (1970) 539–552, doi:10.1016/0020-7683(70)90029-6.

[2] D.A. Saravanos, C.C. Chamis, Unified micromechanics of damping for unidirectional and off-axis fiber composites, J. Compos. Technol. Res. 12 (1) (1990) 31–40, doi:10.1520/ctr10175j.

[3] Z.M. Huang, Simulation of the mechanical properties of fibrous composites by the bridging micromechanics model, Compos. Part A: Appl. Sci. Manuf. 32 (2) (2001) 143–172, doi:10.1016/S1359-835X(00)00142-1.
[4] R.L. Foey, An evaluation of various engineering estimate of the transverse properties of unidirectional composites, Advanced Fibrous Composites, 10 National SAMPE Symposium, 1966.
[5] J.D. Achenbach, H. Zhu, Effect of interphases on micro and macromechanical behavior of hexagonal-array fiber composites, American Society of Mechanical Engineers (Paper), ASME, 1990, https://doi.org/10.1115/1.2897667.
[6] L. Zhang, L.J. Ernst, H.R. Brouwer, Transverse behaviour of a unidirectional composite (glass fibre reinforced unsaturated polyester). Part I. Influence of fibre packing geometry, Mech. Mater. 27 (1) (1998) 13–36, doi:10.1016/S0167-6636(97)00040-9.
[7] A. Wongsto, S. Li, Micromechanical FE analysis of UD fibre-reinforced composites with fibres distributed at random over the transverse cross-section, Compos. Part A: Appl. Sci. Manuf. 36 (9) (2005) 1246–1266, doi:10.1016/j.compositesa.2005.01.010.
[8] G. Karami, M. Garnich, Micromechanical study of thermoelastic behavior of composites with periodic fiber waviness, Compos. Part B: Eng. 36 (3) (2005) 241–248, doi:10.1016/j.compositesb.2004.09.005.
[9] A.A. Gusev, P.J. Hine, I.M. Ward, Fiber packing and elastic properties of a transversely random unidirectional glass/epoxy composite, Compos. Sci. Technol. 60 (4) (2000) 535–541, doi:10.1016/S0266-3538(99)00152-9.
[10] E.E. Ungar, Jr.E. Kerwin, Loss factor of viscoelastic system in terms of energy concepts, J. Acoust. Soc. Am. 34 (1962) 954–958.
[11] G.A. Lesieutre, Modeling frequency-dependent longitudinal dynamic behavior of linear viscoelastic long fiber composites, J. Compos. Mater. 28 (18) (1994) 1770–1782, doi:10.1177/002199839402801802.
[12] A.S. Hadi, J.N. Ashton, Measurement and theoretical modelling of the damping properties of a uni-directional glass/epoxy composite, Compos. Struct. 34 (4) (1996) 381–385, doi:10.1016/0263-8223(96)00005-0.
[13] C.T. Chia, K.A. Khor, Y.W. Gu, F. Boey, Viscoelastic properties of plasma sprayed NiCoCrAlY coatings, Thin Solid Films 405 (1–2) (2002) 146–152, doi:10.1016/S0040-6090(01)01701-1.
[14] J.N. Wei, C.L. Gong, H.F. Cheng, Z.C. Zhou, Z.B. Li, J.P. Shui, F.S. Han, Low-frequency damping behavior of foamed commercially pure aluminum, Mater. Sci. Eng. A 332 (1–2) (2002) 375–381, doi:10.1016/S0921-5093(01)01950-5.
[15] L. Jaouen, B. Brouard, N. Atalla, C. Langlois, A simplified numerical model for a plate backed by a thin foam layer in the low frequency range, J. Sound Vib. 280 (3–5) (2005) 681–698, doi:10.1016/j.jsv.2003.12.038.
[16] W. Flugge, Viscoelasticity, Springer-Verlag Berlin Heidelberg New York, 1967.
[17] R. Chandra, S.P. Singh, and K.A. Gupta, Study on Damping Evaluation in Fiber-Reinforced Composites, India-USA Symposium on Emerging Trends in Vibration and Noise Engineering, The Ohio State University, Columbus, Ohio, December 2001.
[18] R.N. Ronald, R.J. Bankert, E.S. Russell, G.A. Smith, P.K. Wright, Micromechanical Modelling of Fiber/Matrix Interface Effects in Transversely Loaded SiC/Ti-6-4 Metal Matrix Composites, Journal of Composites Technology and Research, 13(1) (1991) 3–13
[19] L. Jungki, H. Han, A. Mal, Effects of anisotropic fiber packing on stresses in composites, Computer Methods in Applied Mechanics and Engineering, 195(33–36), (2006) 4544–4556
[20] L.C., Rogers, C.D. Johnson, D.A. Keinholz, The modal strain energy finite element method and its application to damped laminated beams. Shock Vibr. Bull. (1981) 51.

CHAPTER 5

Modeling of three phase composite

Study of the mechanical behavior of polymer composites is often restricted to a two-phase system. However, with the development of fiber coating technology, the effect of the interface zone, generally called the interphase is of practical importance for the prediction of the overall behavior of the composites. Interphase includes the surface of classical fiber matrix contact as well as a region of finite thickness extending therefrom of which chemical, physical, and morphological features is different from the bulk materials. Interphase provides the composite with improved mechanical performance and structural integrity. The effect of interphase on the mechanical properties of a composite may be investigated by studying two extremes, perfect bonding of fiber, interphase and matrix, and debonding. A mathematical model has been developed for the prediction of elastic moduli of three phase fiber-reinforced composites. This model considers the composite materials consisting of three phases: viz. fiber, interphase and matrix, and incorporates the effect of fiber packing. The Elastic moduli E_{11}, E_{22}, G_{12}, and G_{23} have been evaluated and damping coefficient η_{11}, η_{22}, η_{12}, and η_{23} have been predicted by application of correspondence principle to this developed model. Evaluation of material properties based on finite element method is presented using representative volume element (RVE) approach.

5.1 Mathematical modeling of three phase composite

The two-phase bridging model [1] for unidirectional fiber-reinforced composite has been extended to incorporate the effect of interphase. In two phase model, elastic stresses in matrix material are correlated with the fiber by bridging matrix. The bridging matrix represents the load sharing capacity of one constituent phase in the composite with respect to the other phase. On the same basis stresses in matrix, interphase, and fiber are correlated by Eq. (5.1).

$$\left\{d\sigma_i^m\right\} = \left|A_{ij}^{mI}\right|\left\{d\sigma_j^I\right\}\left\{d\sigma_j^I\right\} = \left|A_{ij}^{If}\right|\left\{d\sigma_j^f\right\} \tag{5.1}$$

Damping in Fiber Reinforced Composite Materials.
DOI: https://doi.org/10.1016/B978-0-323-91186-3.00005-8

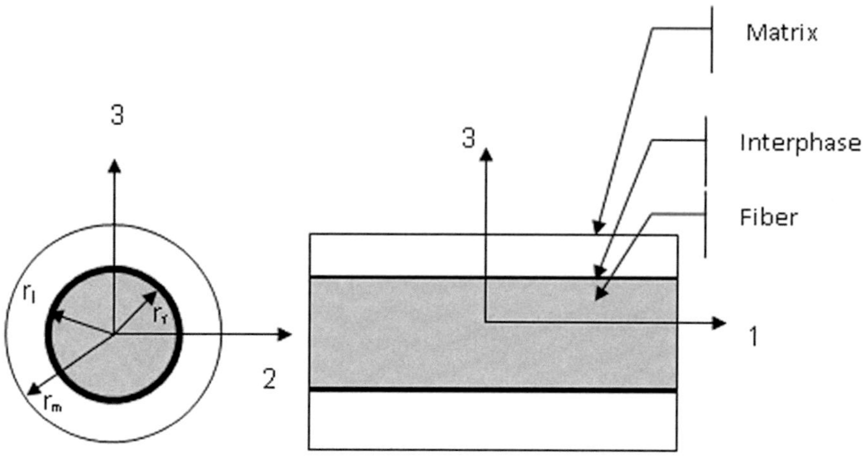

Figure 5.1 Concentric cylinder model for three phase fiber reinforced composite.

Where, $\{d\sigma_i\} = \{d\sigma_{11}, d\sigma_{22}, d\sigma_{33}, d\sigma_{13}, d\sigma_{12}, d\sigma_{23}\}^T$ is the incremental stress tensor.

$[A_{ij}]$ = bridging matrix and superscript f, I and m are fiber, interphase, and matrix respectively. A quantity without suffix represents the composite, subscripts i, j represent the contracted tensorial notation.

Concentric cylinder model for unidirectional three phase fiber-reinforced composite is shown in Fig. 5.1 in the form of RVE. It consists of fiber of radius r_f bounded by interface region of outer radius r_I embedded into matrix of outer radius r_m. Further, condition of perfect fiber-interphase-matrix bonding and variable state of stress is assumed within RVE. Formulation has been carried out for the elastic deformation.

Let V' denote the volume of RVE of unidirectional fiber-reinforced composite as shown in Fig. 5.1.

$$V'_f, V'_I, V'_m = \text{volume of fiber, interphase and matrix in RVE respectively}$$

Volume average stress $\overline{\sigma_i}$ of the composite is defined as:

$$\overline{\sigma_i} = \frac{1}{V'}\int \sigma_i dv = \frac{1}{V'}\left[\int_{V'_f} \sigma_i dv + \int_{V'_i} \sigma_i dv + \int_{V'_m} \sigma_i dv\right] \tag{5.2}$$

$$\bar{\sigma}_i = \left(\frac{V'_f}{V'}\right)\left(\frac{1}{V'_f}\int_{v'_f} \sigma_i dv\right) + \left(\frac{V'_I}{V'}\right)\left(\frac{1}{V'_I}\int_{v'_i} \sigma_i dv\right) + \left(\frac{V'_m}{V'}\right)\left(\frac{1}{V'_m}\int_{v'_m} \sigma_i dv\right)$$

$$\bar{\sigma}_i = V_f\overline{\sigma_i^f} + V_I\overline{\sigma_i^I} + V_m\overline{\sigma_i^m} \tag{5.3}$$

Where,

$$\overline{\sigma_i^f} = \left[\frac{1}{V'_f}\int_{v'_f}\sigma_i dv\right],\ \overline{\sigma_i^I} = \left[\frac{1}{V'_I}\int_{v'_I}\sigma_i dv\right],\ \overline{\sigma_i^m} = \left[\frac{1}{V'_m}\int_{v'_m}\sigma_i dv\right],\ V_f = \frac{V'_f}{V'},$$
$$V_I = \frac{V'_I}{V'},\ V_m = \frac{V'_m}{V'}$$

where V_f, V_I, and V_m represents volume fraction of fiber, interphase, and matrix respectively and

$\overline{\sigma_i^f}$, $\overline{\sigma_i^I}$ and $\overline{\sigma_i^m}$ are volume average constituent stresses. Similarly, for volume averaged strain is given as

$$\bar{\varepsilon}_i = V_f\overline{\varepsilon_i^f} + V_I\overline{\varepsilon_i^I} + V_m\overline{\varepsilon_i^m} \tag{5.4}$$

In further discussion volume averaged stresses and strains, will be represented by omitting the over bar. Volume averages stresses and strains in RVE of lamina in incremental form are given by Eqs. (5.5) and (5.6).

$$\{d\sigma\} = V_f\{d\sigma^f\} + V_I\{d\sigma^I\} + V_m\{d\sigma^m\} \tag{5.5}$$

And

$$\{d\varepsilon\} = V_f\{d\varepsilon^f\} + V_I\{d\varepsilon^I\} + V_m\{d\varepsilon^m\} \tag{5.6}$$

Constitutive equations correlating the averaged stresses and strains in different phases of the RVE are expressed as

$$\{d\varepsilon^f\} = [S^f]\{d\sigma^f\} \tag{5.7}$$

$$\{d\varepsilon^I\} = [S^I]\{d\sigma^I\} \tag{5.8}$$

$$\{d\varepsilon^m\} = [S^m]\{d\sigma^m\} \tag{5.9}$$

Substituting Eqs. (5.7), (5.8), and (5.9) into Eq. (5.6) and after simple algebraic manipulation we get

$$\{d\varepsilon\} = [S_{ij}]\left[V_f + V_I\left[A_{ij}^{If}\right] + V_m\left[A_{ij}^{mI}\right]\left[A_{ij}^{If}\right]\right]\{d\sigma^f\} \tag{5.10}$$

Where

$$S_{ij} = \left[V_f S_{ij}^f + V_I S_{ij}^I A_{ij}^{If} + V_m S_{ij}^m A_{ij}^{mI} A_{ij}^{If}\right]\left[V_f[I] + V_I A_{ij}^{If} + V_m A_{ij}^{mI} A_{ij}^{If}\right]^{-1} \tag{5.11}$$

Overall compliance matrix of lamina is given by Eq. (5.12)

$$S_{ij} = \begin{bmatrix} [S_{ij}]_\sigma & 0 \\ 0 & [S_{ij}]_\tau \end{bmatrix} \tag{5.12}$$

where $[S_{ij}]_\sigma$ and $[S_{ij}]_\tau$ are normal and shear compliance matrices

$$[S_{ij}]_\sigma = \begin{bmatrix} 1/E_{11} & -\nu_{12}/E_{11} & -\nu_{12}/E_{11} \\ & 1/E_{22} & -\nu_{23}/E_{22} \\ \text{symmetry} & & 1/E_{22} \end{bmatrix} \tag{5.13}$$

$$[S_{ij}]_\tau = \begin{bmatrix} 1/G_{23} & 0 & 0 \\ 0 & 1/G_{12} & 0 \\ 0 & 0 & 1/G_{12} \end{bmatrix}$$

Material parameters E_{22}, G_{23}, and ν_{23} are related to each other with as in Eq. (5.14)

$$G_{23} = \frac{E_{22}}{2(1 + \nu_{23})} \tag{5.14}$$

Thus, only five independent elastic moduli exist. Hence, only five independent elements in bridging matrix are required. General form of bridging matrix is given by Eq. (5.15),

$$[A_{ij}] = \begin{bmatrix} a_{11} & a_{12} & a_{13} & 0 & 0 & 0 \\ a_{21} & a_{22} & a_{23} & 0 & 0 & 0 \\ a_{31} & a_{32} & a_{33} & 0 & 0 & 0 \\ 0 & 0 & 0 & a_{44} & 0 & 0 \\ 0 & 0 & 0 & 0 & a_{55} & 0 \\ 0 & 0 & 0 & 0 & 0 & a_{66} \end{bmatrix} \tag{5.15}$$

we take $a_{11}, a_{22}, a_{12}, a_{23}, \textit{and } a_{66}$ to be independent.

Bridging matrix between fiber and interphase is represented as

$$[A_{ij}^{fI}] = \begin{bmatrix} a_{11}^{fI} & a_{12}^{fI} & a_{13}^{fI} & 0 & 0 & 0 \\ a_{21}^{fI} & a_{22}^{fI} & a_{23}^{fI} & 0 & 0 & 0 \\ a_{31}^{fI} & a_{32}^{fI} & a_{33}^{fI} & 0 & 0 & 0 \\ 0 & 0 & 0 & a_{44}^{fI} & 0 & 0 \\ 0 & 0 & 0 & 0 & a_{55}^{fI} & 0 \\ 0 & 0 & 0 & 0 & 0 & a_{66}^{fI} \end{bmatrix} \tag{5.16a}$$

whereas matrix between interphase and matrix is represented as

$$\left[A_{ij}^{\mathrm{Im}}\right] = \begin{bmatrix} a_{11}^{\mathrm{Im}} & a_{12}^{\mathrm{Im}} & a_{13}^{\mathrm{Im}} & 0 & 0 & 0 \\ a_{21}^{\mathrm{Im}} & a_{22}^{\mathrm{Im}} & a_{23}^{\mathrm{Im}} & 0 & 0 & 0 \\ a_{31}^{\mathrm{Im}} & a_{32}^{\mathrm{Im}} & a_{33}^{\mathrm{Im}} & 0 & 0 & 0 \\ 0 & 0 & 0 & a_{44}^{\mathrm{Im}} & 0 & 0 \\ 0 & 0 & 0 & 0 & a_{55}^{\mathrm{Im}} & 0 \\ 0 & 0 & 0 & 0 & 0 & a_{66}^{\mathrm{Im}} \end{bmatrix} \tag{5.16b}$$

The independent elements a_{11}, $a_{22}, a_{12}, and\ a_{23}$ depend on the elastic properties of the matrix, interphase, and fiber and also on fiber packing.

Due to the axisymmetry of concentric cylinder

$$a_{33} = a_{22} \tag{5.17a}$$

$$a_{13}^{fI} = a_{12}^{fI} = \frac{\left(S_{12}^{f} - S_{12}^{I}\right)\left(a_{11}^{fI} - a_{22}^{fI}\right)}{\left(S_{11}^{f} - S_{11}^{I}\right)} \tag{5.17b}$$

$$a_{13}^{\mathrm{Im}} = a_{12}^{\mathrm{Im}} = \frac{\left(S_{12}^{I} - S_{12}^{m}\right)\left(a_{11}^{\mathrm{Im}} - a_{22}^{\mathrm{Im}}\right)}{\left(S_{11}^{I} - S_{11}^{m}\right)} \tag{5.17c}$$

$$a_{11}^{fI} = E^{I}/E^{f},\ a_{11}^{\mathrm{Im}} = E^{m}/E^{I} \tag{5.17d}$$

$$a_{22}^{fI} = \beta + (1 - \beta)E^{I}/E^{f} \tag{5.17e}$$

$$a_{22}^{\mathrm{Im}} = \beta + (1 - \beta)E^{m}/E^{I} \tag{5.17f}$$

$$a_{44}^{fI} = \beta + (1 - \beta)E^{I}/E^{f} \tag{5.17g}$$

$$a_{44}^{\mathrm{Im}} = \beta + (1 - \beta)E^{m}/E^{I} \tag{5.17h}$$

$$a_{55}^{fI} = a_{66}^{fI} = \alpha + (1 - \alpha)G^{I}/G^{f} \tag{5.17i}$$

$$a_{55}^{\mathrm{Im}} = a_{66}^{\mathrm{Im}} = \alpha + (1 - \alpha)G^{m}/G^{I} \tag{5.17j}$$

Where α and β are the longitudinal and transverse fiber packing factor.

Using the bridging compliance matrices $[S_{ij}]$, four elastic moduli E_{11}, E_{22}, G_{12}, and G_{23} of unidirectional fiber composite are derived.

$$E_{11} = \frac{V_f + V_I a_{11fI} + V_m a_{11fI} a_{11\mathrm{Im}}}{V_f S_{11f} + V_I S_{11I} a_{11fI} + V_m S_{11m} a_{11fI} a_{1\mathrm{Im}}} \tag{5.18a}$$

$$E_{22} = \frac{\left(V_f + V_I a_{22fI} a_{22\mathrm{Im}}\right)\left(V_f + V_I a_{11fI} + V_m a_{11fI} a_{11\mathrm{Im}}\right)}{V_f^2 S_{22f} + a_{22fI} a_{11fI}\left(V_I^2 S_{22I} + V_m^2 S_{22m} a_{11fI} a_{22\mathrm{Im}}\right) - V_f V_I A - V_I V_m B - V_m V_f C}$$

where,

$$A = a_{12fI}\left(S_{12f} - S_{12I}\right) - S_{22f}a_{11fI} - S_{22I}a_{22fI}$$

$$B = a_{11fI}(S_{12I} - S_{12m})\left(a_{11fI}a_{12\mathrm{Im}} - a_{12fI}a_{11\mathrm{Im}} + a_{22\mathrm{Im}}a_{12fI}\right)$$

$$C = a_{12fi}a_{22\mathrm{Im}}\left(S_{12f} - S_{12m}\right) + a_{11fI}\left(S_{12f}a_{12\mathrm{Im}} + S_{22f}a_{11\mathrm{Im}} - S_{12m}a_{12\mathrm{Im}}\right) \tag{5.18b}$$

$$G_{12} = \frac{G_f G_I G_m (V_f + V_I a_{66fI} + V_m a_{66fI} a_{66\mathrm{Im}}}{G_f G_m V_I a_{66fI} + G_f G_I V_m a_{66\mathrm{Im}} + G_I G_m V_f} \tag{5.18c}$$

$$G_{23} = \frac{V_f + V_I a_{44fI} + V_m a_{44fI} a_{44\mathrm{Im}}}{V_f S_{44f} + V_I S_{44I} a_{44fI} + V_4 S_{44m} a_{44fI} a_{44\mathrm{Im}}} \tag{5.18d}$$

Further, using the correspondence principle in which E_f, E_I, E_m becomes complex with the loss modulus associated as the imaginary part. Thus, the compliance matrix would be complex. As a result, composite moduli in complex form are obtained as given in Eq. (5.19) representing the storage moduli (E_{ij}') and loss moduli (E_{ij}'') components as real and imaginary parts respectively. Using these loss factors $\eta_{11}, \eta_{22}, \eta_{12}$, and η_{23} are predicted.

$$E_{ij} = E_{ij}' + jE''_{ij},\ n_{ij} = \frac{E''_{ij}}{E_{ij}'} \tag{5.19}$$

5.2 Finite elements model

The 3-D finite element model with varying values of fiber volume fraction, interphase volume fraction, and for various loading with single cell square array packing are analyzed. Fig. 5.2 shows a typical 3-D finite element model for transverse loading modeled in NISA version 15 finite element software. The model consists of 1-2560 eight nodded hexahedron elements in fiber, 2561-3840 eight nodded hexahedron elements in interphase and 3840-5960 eight nodded hexahedron elements in matrix as shown in Fig. 5.2.

Applying the strain energy concept, the unit cell [2] is analyzed to evaluate the damping capacity of composite from those of its constituents.

Loss factor is expressed as

$$\eta = \frac{\sum_{i=1}^{n} \eta_i W_i}{\sum_{i=1}^{n} W_i} \tag{5.20}$$

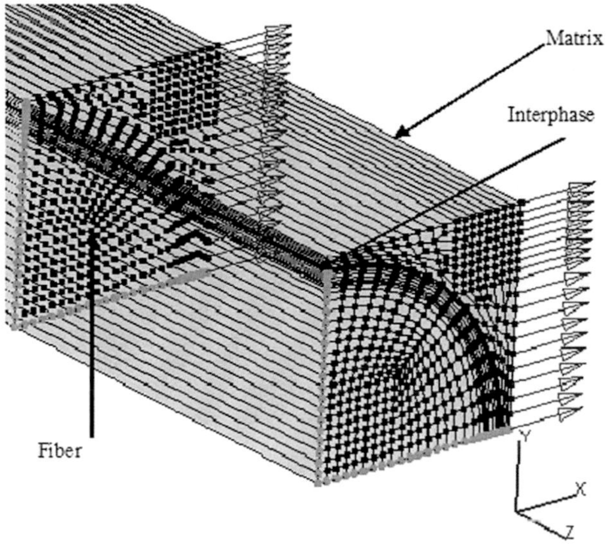

Figure 5.2 Finite element model of Three phase fiber-reinforced composite. Number of elements $N_f = 2560$, $N_I = 2561\text{-}3840$, $N_m = 3841\text{-}5960$, $V_f = 0.6$, $V_I = 0.1$.

The loss factor for three phase model considering the effect of interphase is expressed as

$$\eta = \frac{\eta_f W_f + \eta_I W_I + \eta_m W_m}{W_f + W_I + W_m} \tag{5.21}$$

The strain energy stored in composite under loading is written as

$$W_c = \frac{1}{2}\int_V \sigma_{ij}\varepsilon_{ij}dV = \frac{1}{2}\int_V \sigma_{ij}^T S_{ij}\sigma_{ij}dV \tag{5.22}$$

The strain energy stored in fiber, interphase, and matrix in a unit cell is written as

$$\begin{aligned} W_f &= \frac{1}{2}\sum \left\{\sigma_{ij}^T\right\}_f \{S_{ij}\}_f \{\sigma_{ij}\}_f dV_f \\ W_I &= \frac{1}{2}\sum \left\{\sigma_{ij}^T\right\}_I \{S_{ij}\}_I \{\sigma_{ij}\}_I dV_I \\ W_m &= \frac{1}{2}\sum \left\{\sigma_{ij}^T\right\}_m \{S_{ij}\}_m \{\sigma_{ij}\}_m dV_m \end{aligned} \tag{5.23}$$

Equating expression W_c as sum of the strain energy stored in the components, we can evaluate the overall elastic moduli. These results for moduli and loss factors are compared with output of the bridging model for three phase composites and discussed in details for various parameters of the composites.

Table 5.1 Constituent material properties used as input for bridging model.

Properties	E-glass fiber	Epoxy matrix	Strong interphase	Weak interphase
E (GPa)	72.4	2.76	37.58	0.5
G (GPa)	30.2	1.02	15.61	0.178
υ	0.2	0.35	0.204	0.4
η	0.0018	0.015	0.0084	0.0084

5.3 Frequency dependence of three phase composite

Frequency is one of the important parameters for the study of the dynamic behavior of fiber-reinforced composite materials. In a three-phase composite, matrix is assumed as a viscoelastic material, whereas the fiber is assumed as an elastic material for evaluation of its elastic modulus and loss factors. Viscoelastic material is modeled by the four parameter Maxwell and Voigt model as discussed in Chapter 4. Formulation of four elastic modulus as functions of frequency has been obtained using Eqs. (4.49–4.50) and Eqs. (5.18). By the application of viscoelastic correspondence, principal loss factors are predicted as functions of frequency. The effects of longitudinal and transverse fiber packing factors on loss factors have also been evaluated with respect to frequency.

5.4 Results and discussion

Elastic moduli and loss factors are predicted using a developed three-phase bridging model for various values of the fiber volume fraction with constituent properties given in Table 5.1. The effect of fiber packing factors is evaluated on the elastic moduli and loss factors. The effect of strong interphase thickness and weak interphase thickness has been evaluated on loss factor. Strong interphase has a higher elastic modulus than the elastic modulus of the matrix and the weak interphase's elastic modulus is less than the elastic modulus of the matrix as given in Table 5.1. Finite element/strain energy approach is also used to predict elastic moduli (E_{11}, E_{22}, G_{12}, and G_{23}) and loss factors ($\eta_{11}, \eta_{22}, \eta_{12}$, and η_{23}).

5.4.1 Strain energy variation using FEM

Fiber reinforced composite materials having three-phase namely, fiber, interphase, and matrix have been modeled in finite element software NISA 15. Hexahedron element (NKTP = 4) is taken for meshing the RVE of the composite. Appropriate constituent properties and boundary

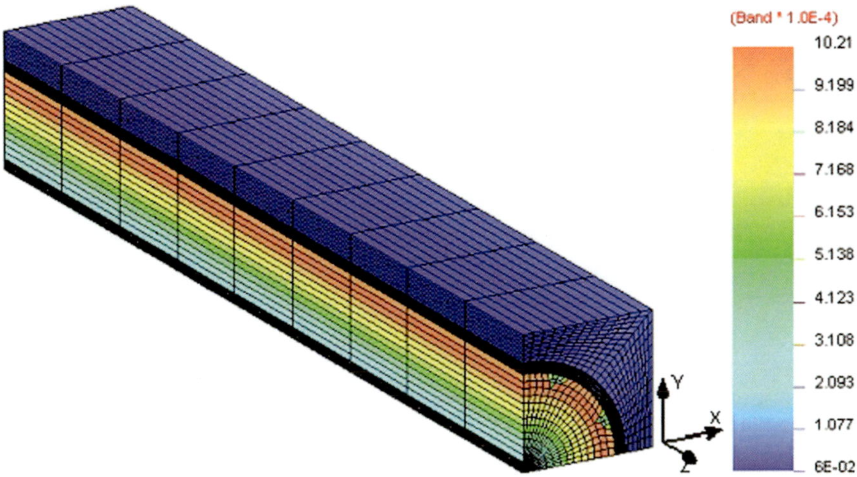

Figure 5.3 Strain energy variation for longitudinal loading for strong interphase, $V_f = 0.4$ and $V_i = 0.1$.

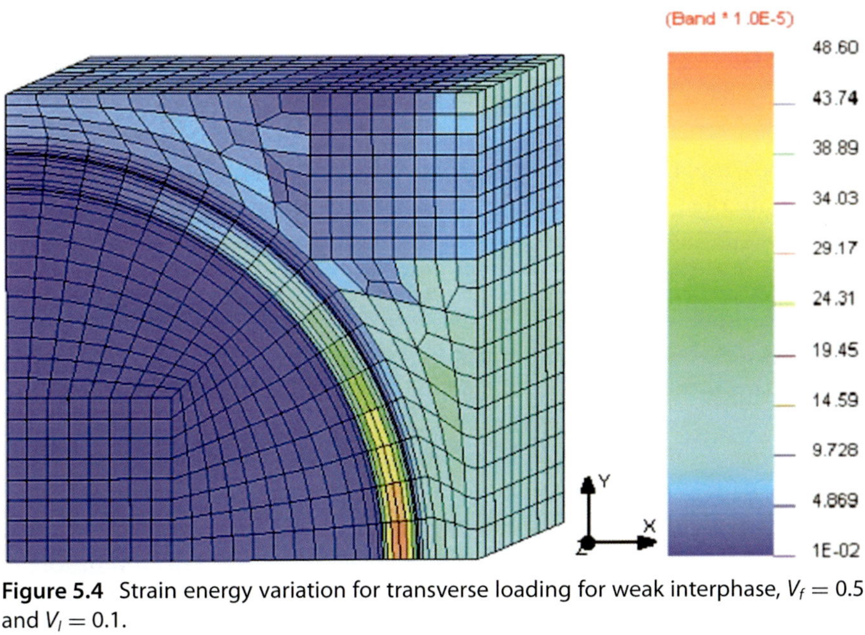

Figure 5.4 Strain energy variation for transverse loading for weak interphase, $V_f = 0.5$ and $V_i = 0.1$.

conditions have been applied to the 3D finite element model under different loadings (Table 4.2). Further, the loss factors are predicted using Eq. (5.21). Strain energy variation within the constituent of the square cell RVE under different loading is shown in Fig. 5.3–5.6. Variation of strain energy in longitudinal direction is shown in Fig. 5.3 for 40% fiber volume fraction

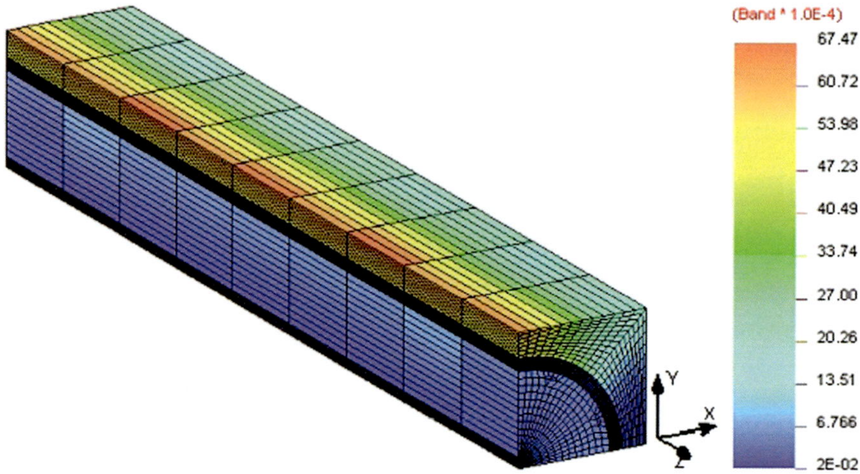

Figure 5.5 Strain energy variation for longitudinal shear loading for strong interphase, $V_f = 0.4$ and $V_I = 0.1$.

and 0.1 micron thick interphase. Strain energy variation in fiber is more in comparison to the matrix and interphase. Fig. 5.4 shows the variation of strain energy in fiber, interphase, and matrix for transverse loading considering weak interphase. Strain energy remains constant for the fiber region because the strain produced is very small and remains constant throughout the fiber. Interphase strain energy variation is maximum as interphase modulus is low hence strain is very high. Strain energy variation is there, but with less magnitude. Fig. 5.5 depicts the variation of strain energy for longitudinal shear loading for strong interphase. Strain energy variation is more in the matrix and variation is almost constant in fiber and interphase. As elastic modulus of interphase is more than matrix and hence variation is very less. Fig. 5.6 elaborates the variation of strain energy for transverse shear loading with weak interphase. Strain energy variation is more in interphase and in matrix.

Figs. 5.7 and 5.8 demonstrate the variation of normal elastic moduli E_{11} and E_{22} with respect to fiber volume fraction V_f. The longitudinal elastic modulus E_{11} shows insensitiveness to the fiber packing factors α and β. In this case, fiber, interphase and matrix are stressed in longitudinal direction and not much interaction takes place between fiber and matrix. Therefore, arrangement of fiber and matrix does not have any role in taking the load. E_{11} as predicted by all models bridging, and concentric cylinder assembly method (CCA) [3] and Mori-Tanaka is practically same and varies linearly with fiber volume fraction. Fig. 5.8 indicates the dependence

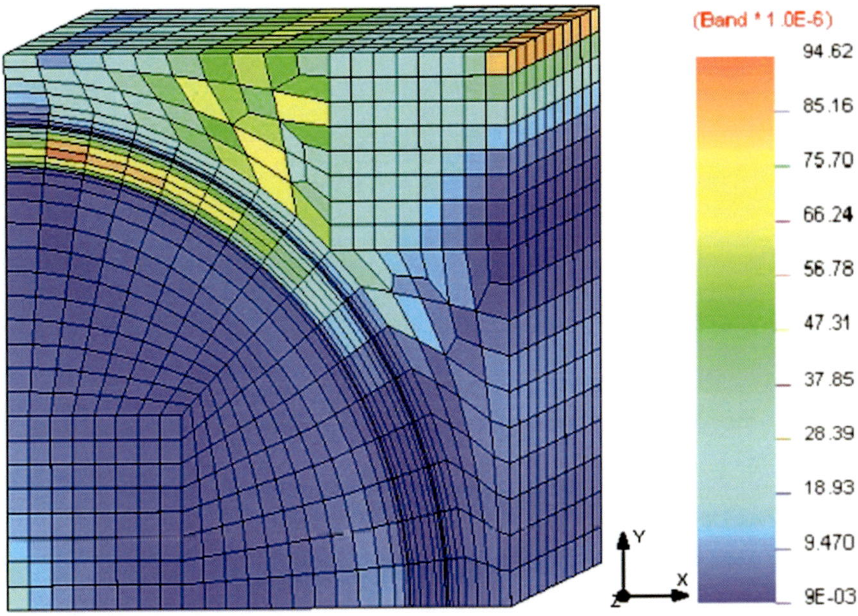

Figure 5.6 Strain energy variation for transverse shear loading for weak interphase, $V_f = 0.5$ and $V_I = 0.1$.

of transverse elastic modulus E_{22} on transverse packing factor β. Elastic modulus E_{22} for GFRE shows an increasing trend as the fiber volume fraction for all values of transverse fiber packing factor in the range of 0.5 to 1. However, for a particular fiber volume fraction, elastic modulus decreases as the transverse fiber packing factor, β increases from 0.5 to 1. The fiber packing effect is more pronounced at higher values of fiber volume fraction. At $V_f = 0.4$, transverse elastic modulus E_{22} decreases by 36% whereas at $V_f = 0.6$, E_{22} decreases by 42% as β increases from 0.5 to 1.0. Fig. 5.8 also shows the comparison with the CCA method. For $\beta = 1$ bridging model results match with those obtained with the CCA method. FEM results are in good agreement with the bridging model results for lower value of fiber volume fractions (up to 50%).

The bridging model results predicted for transverse elastic moduli are compared with finite element results reported by Wacker et al. [4] for the following input data.

$$E_f = 84\,\text{GPa},\ \nu_f = 0.22$$
$$E_m = 4\,\text{GPa},\ \nu_m = \nu_i = 0.34$$
$$r_f = 8.5\mu\,\text{m and interphase thickness} = 1\mu m$$

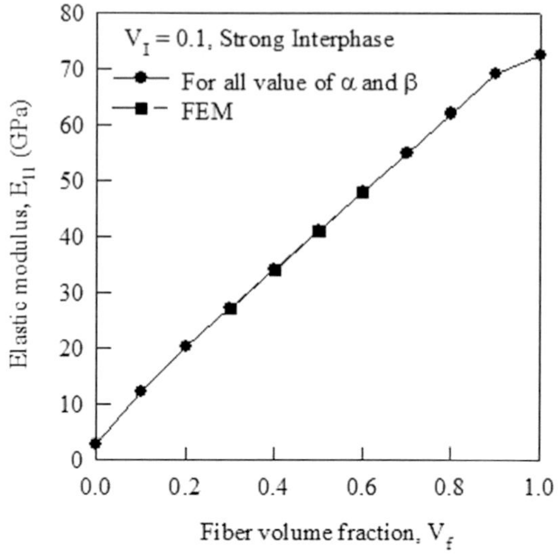

Figure 5.7 Variation of Elastic modulus E_{11} with respect to fiber volume fraction.

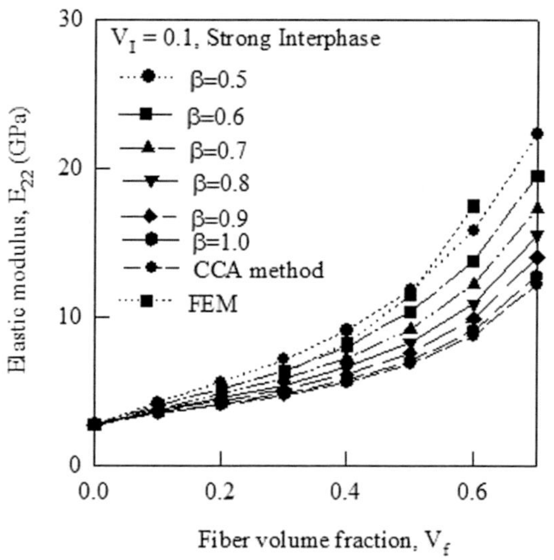

Figure 5.8 Variation of Elastic modulus E_{22} with respect to fiber volume fraction.

The results of the bridging model for transverse fiber packing factor $\beta = 0.51$ show very close matching to the FEM results as seen from Table 5.2.

Fig. 5.9 and 5.10 show the effect of fiber volume fraction V_f on shear moduli G_{12} and G_{23} respectively. In Fig. 5.9 various curves represent variation of G_{12} for fiber packing factor $0.5 < \alpha < 1.0$. With the increase

Table 5.2 Transverse elastic modulus E_{22} for unidirectional GFRE.

V_f	Fiber packing factor β	A	*Ei* (GPa)	Three phase bridging model E_{22} (GPa)	FEM Wacker et al. [4] E_{22} (GPa)
0.5	0.51	0.5	6	13.46	13.71
0.5	0.51	0.5	8	14.68	14.68
0.5	0.51	0.5	12	16.10	15.91

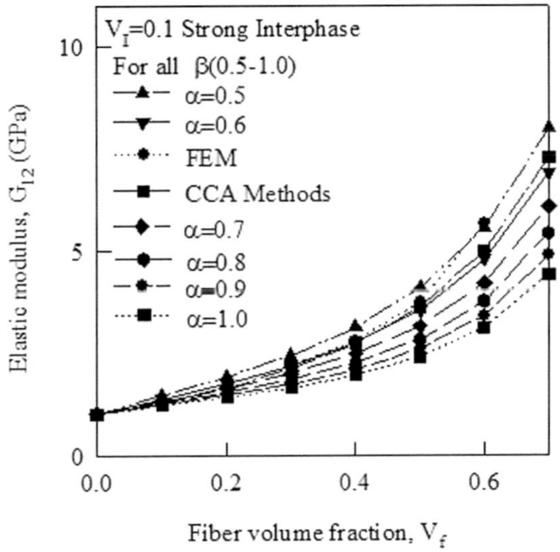

Figure 5.9 Variation of Elastic modulus G_{12} with respect to fiber volume fraction.

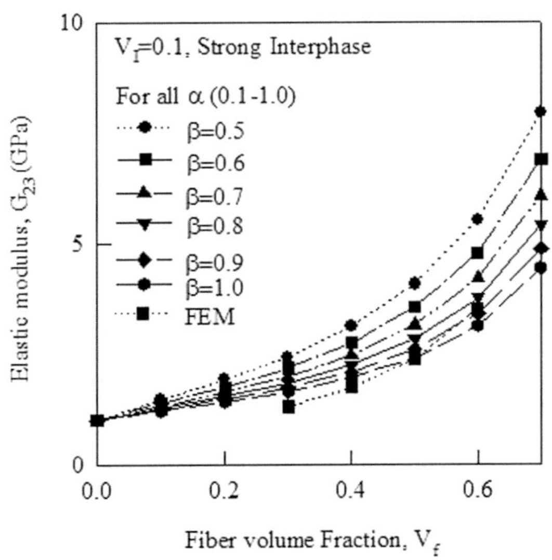

Figure 5.10 Variation of Elastic modulus G_{23} with respect to fiber volume fraction.

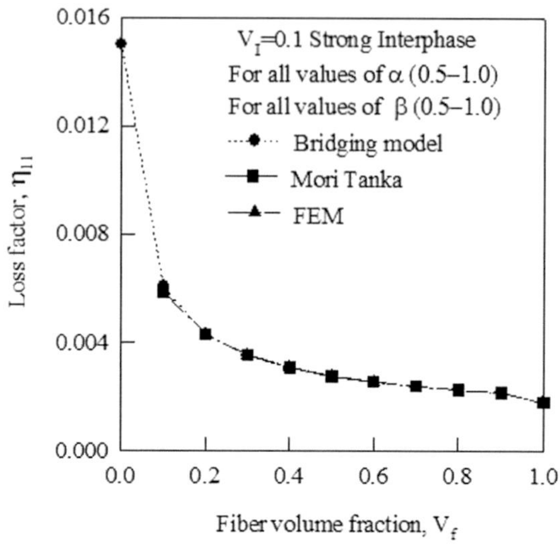

Figure 5.11 Variation of loss factor η_{11} with respect to fiber volume fraction.

in the value of α, G_{12} decreases rapidly for lower range of α. Increase in longitudinal shear modulus is more for lower value of α when fiber volume fraction is large. The results for G_{12} indicate that the transverse packing factor β has no effect on its prediction. FEM results are closer to bridging model results for $\alpha = 0.6$ for lower value of fiber volume fraction. Fig. 5.10 shows similar trend for transverse shear modulus, G_{23} as in case of longitudinal shear modulus, G_{12} but with different magnitudes, in relation to different values of transverse packing factor β. Further, the magnitude of G_{23} is independent of longitudinal packing factor α.

Figs. 5.11 and 5.12 illustrate the effects of V_f on the longitudinal normal loss factor (η_{11}) and the longitudinal shear loss factor (η_{12}) respectively. In Fig. 5.11, the longitudinal loss factor, η_{11} decreases with the increase of fiber volume fraction. Decrease in magnitude of loss factors is quite rapid for lower value of V_f and it shows insensitiveness towards both the fiber packing factors α and β. The predictions for η_{11} by FEM with a single cell square array packing and Mori Tanka shows very good agreement with the bridging model results for three phase GFRE composite. Fig. 5.12 shows that for fiber packing factor $\alpha = 0.7$ loss factor η_{12} is very near to the results obtained by CCA methods. Longitudinal shear loss factor as obtained by bridging model compare very well with CCA method for all value of V_f. FEM results of loss factor η_{12} are comparable with the bridging model results for longitudinal fiber packing factor at $\alpha = 0.7$ indicating its effectiveness to account for variable state of stress in the composites.

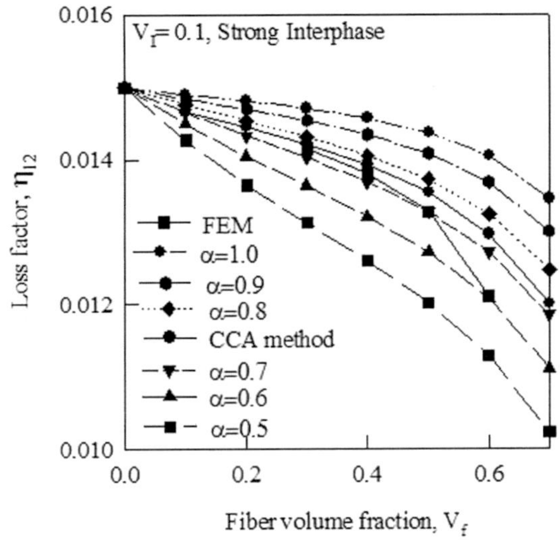

Figure 5.12 Variation of loss factor η_{12} with respect to fiber volume fraction.

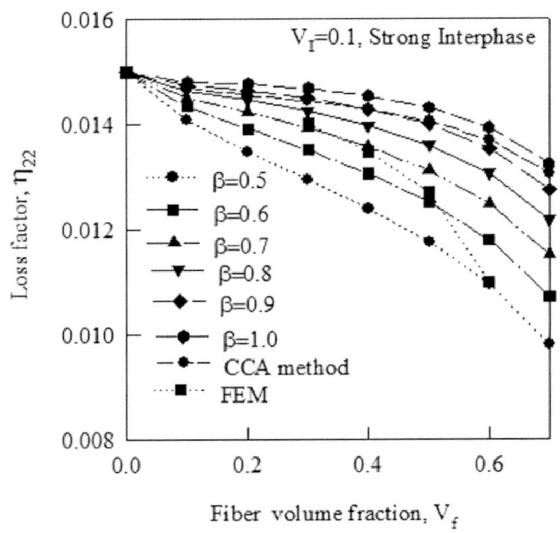

Figure 5.13 Variation of transverse loss factor η_{22} with respect to fiber volume fraction.

Figs. 5.13 and 5.14 demonstrate the variation of transverse normal loss factor η_{22} and transverse shear loss factor η_{23} with respect to fiber volume fraction V_f. Loss factors η_{22} and η_{23} both decrease with increase in V_f however their magnitude increases with the increase of transverse fiber packing factor β. Fig. 5.13 shows that with the increase of β loss factor η_{22} increases rapidly at higher value of V_f. At $V_f = 0.4$ increase in η_{22} is

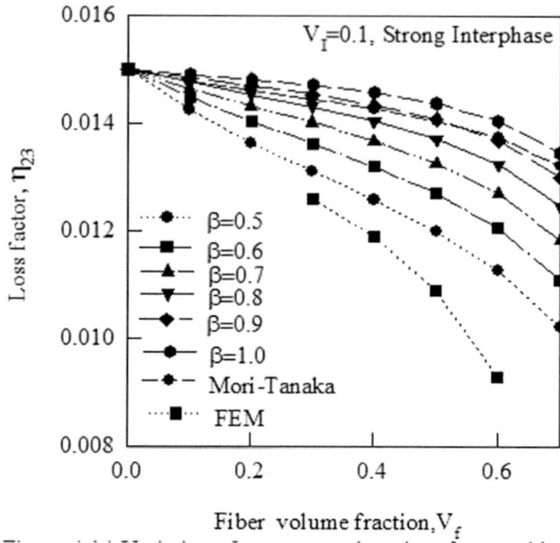

Figure 5.14 Variation of transverse shear loss factor η_{23} with respect to fiber volume fraction.

12.7% for increase in β from 0.5 to 0.8 and at $V_f = 0.6$ increase in loss factor η_{22} is 19.2% for the same value of β. For $\beta = 0.9$ loss factor η_{22} results are almost similar to the one obtained by CCA method result. Prediction for loss factor, η_{22} obtained for square cell FEM compares well with that of three phase bridging model for transverse fiber packing factor, $\beta = 0.7$. Variation of transverse shear loss factor η_{23} with respect to fiber volume fraction for different transverse fiber packing factor is shown in Fig. 5.14. It is observed that for higher value of β loss factor decreases slowly as compared to the one at lower value of β. As V_f increases from 0.4 to 0.6 loss factor η_{23} decreases by 10.4% for $\beta = 0.5$, and 3.6% for $\beta = 1.0$. The prediction for η_{23} by FEM with single cell square array packing shows lower results.

The loss factors η_{11} and η_{12} of fiber-reinforced composite evaluated by varying the strong interphase volume fraction from 0.02 to 0.1 are shown in Fig. 5.15 and 5.16. Fig. 5.15 shows that loss factor, η_{11} increases with the increase in interphase volume fraction. Moreover η_{11} is independent of fiber packing factor α and β and results match well with those from FEM/strain energy approach. The loss factor η_{12} in Fig. 5.16 illustrates its dependence on α which increases with increase in longitudinal fiber packing factor α. However, it decreases marginally with increase in interphase volume fraction V_i. Loss factor η_{12} for composite predicted using FEM and the bridging model are compared in Fig. 5.16. Further results from FEM and the Bbridging model for $\beta = 0.9$ show good agreement with each other.

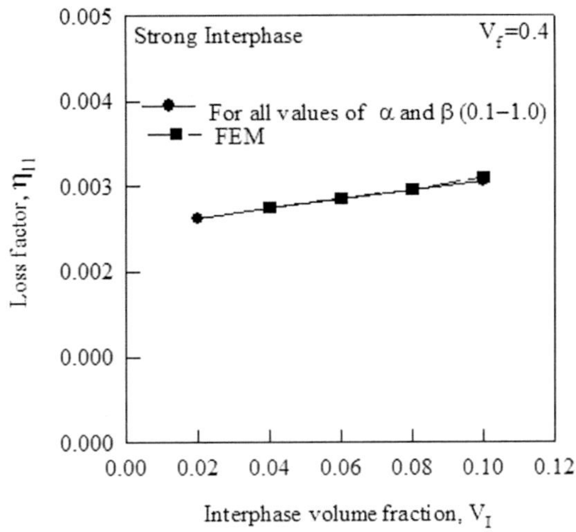

Figure 5.15 Effect of interphase volume fraction on η_{11}.

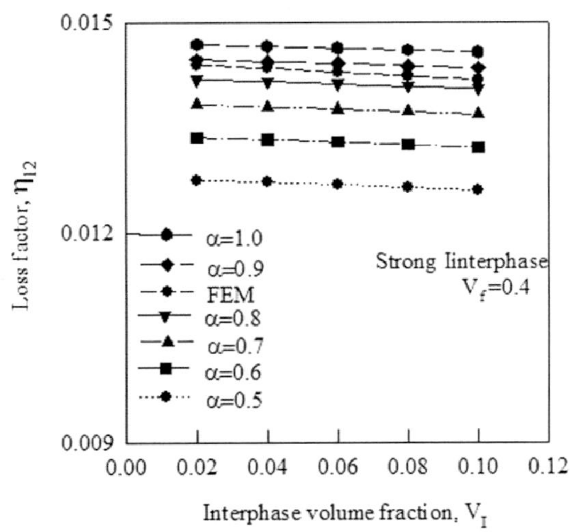

Figure 5.16 Effect of interphase volume fraction on η_{12}.

Figs. 5.17 and 5.18 demonstrate the variation of the transverse normal loss factor η_{22} and the transverse shear loss factor η_{23} with respect to changes in strong interphase volume fraction respectively. Different curves represent behavior of η_{22} and η_{23} for different values of β. The loss factor η_{22} is compared with FEM and Vantomme [5] results in Fig. 5.17. Results from the bridging model are comparable to FEM results for single cell models. For $\beta = 1.0$ bridging model results match well with Vantomme model results

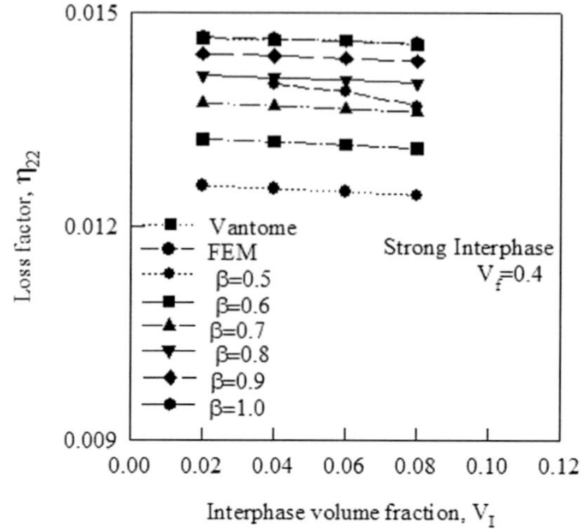

Figure 5.17 Effect of interphase volume fraction on η_{22}.

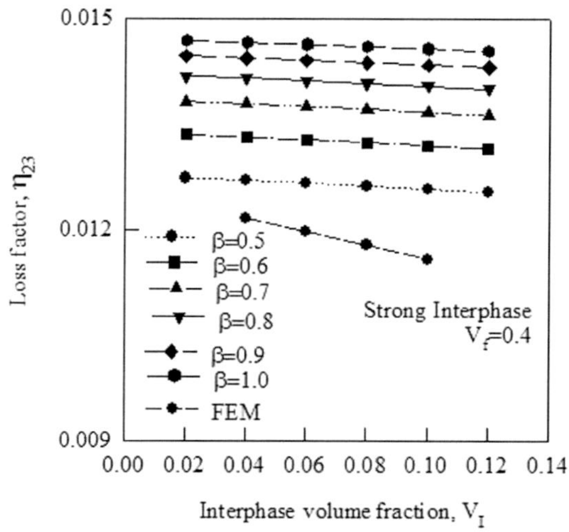

Figure 5.18 Effect of interphase volume fraction on η_{23}.

where the effect of fiber packing is not considered. The Vantomme model is based upon the law of mixture, considering the uniform state of stress in fiber, interphase, and matrix. This indicates that $\beta = 1$ refers to the model with a single fiber which does not consider the effect of fiber packing. Square fiber packing has been considered in the FEM model for the prediction of η_{22}. In present model, the effect of fiber packing is obtained by variation of β. Fig. 5.18 illustrates the variation of η_{23} with strong interphase.

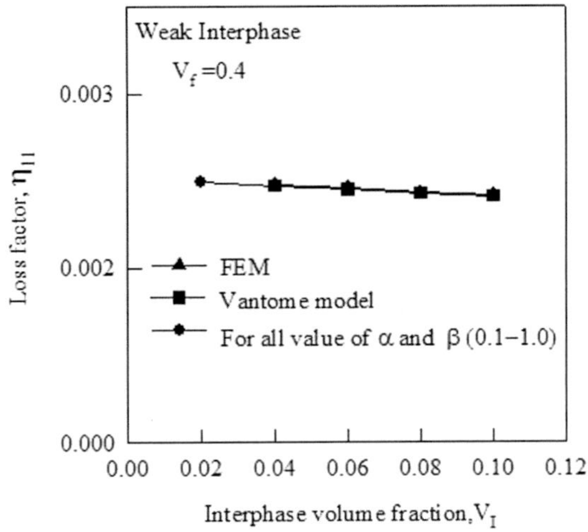

Figure 5.19 Effect of interphase volume fraction on η_{11}.

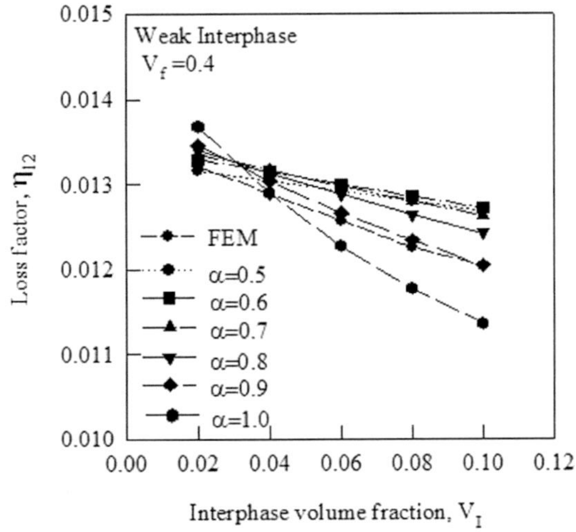

Figure 5.20 Effect of interphase volume fraction on η_{12}.

With an increase in β loss factor η_{23} increases, but the increase in η_{23} is not much for higher value of β.

Variation of η_{11} and η_{12} with respect to interphase volume fraction V_i is presented in Figs. 5.19 and 5.20 considering weak interphase for 40% fiber volume fraction. The loss factor η_{11} decreases marginally with increase in V_i. Fiber packing factor α and β has no effect on η_{11}. Fig. 5.19 also shows comparison with FEM [6] results. Both results have same trend with almost

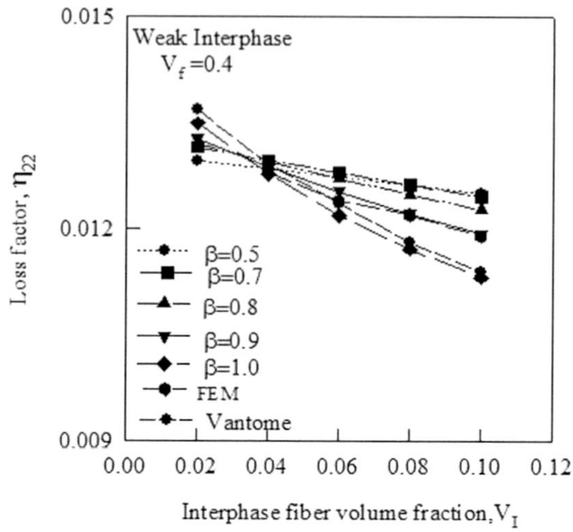

Figure 5.21 Effect of interphase volume fraction on η_{22}.

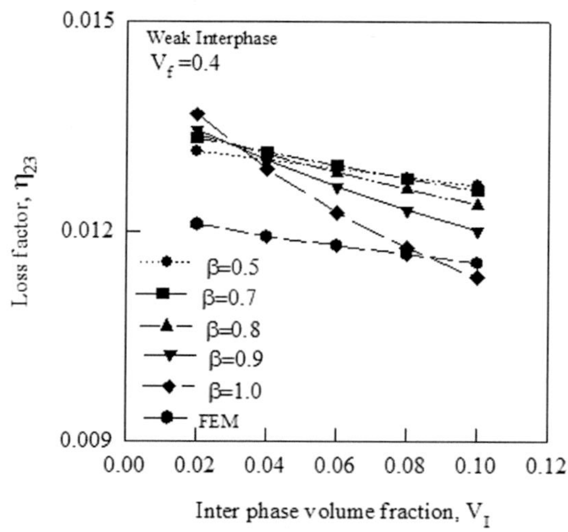

Figure 5.22 Effect of interphase volume fraction on η_{23}.

same magnitude. The bridging model results are in good agreement with Vantomme results. Fig. 5.20 shows decrease in η_{12} with increase in V_I. The loss factor, η_{12} is more for higher value of α. For $\alpha = 0.9$, bridging model results are in good agreement with the FEM results for higher value of interphase volume fraction ($V_I = 0.06$–0.08).

Figs. 5.21 and 5.22 show the variation of η_{22} and η_{23} with respect to weak interphase volume fraction V_I. Fig. 5.21 illustrates that the loss

Table 5.3 Fiber packing factor for three phase GFRE composite.

V_f	FEM result	Proposed bridging model result	Optimized fiber packing factor
	Transverse shear loss factor (η_{23})		
0.4	0.0142	0.014	$\beta = 0.77$
0.5	0.0138	0.0138	
0.6	0.0133	0.0133	
0.7	0.0122	0.0121	
	Longitudinal shear loss factor (η_{12})		
0.4	0.0126	0.0121	$\alpha = 0.44$
0.5	0.0119	0.0115	
0.6	0.0109	0.0107	
0.7	0.0093	0.0096	

factor η_{22} predicted by the bridging model for a higher value of transverse fiber packing factor β that is, $\beta = 1$ it decreases rapidly with increase in V_I, whereas for lower value of β from 0.5 to 0.8 the variation of η_{22} is comparatively very small. Further for lower value of β the loss factor η_{22} is higher as compared to the one when $\beta = 1$ above interphase volume fraction $V_i = 0.04$. Loss factor η_{22} increases with decrease in value of β from 0.9 to 0.5 for interphase volume fraction $V_I = 0.04$. Fig. 5.21 also depicts the comparison of bridging model results with FEM and Vantomme model results. For $\beta = 1$ bridging model results are in good agreement with Vantomme model for interphase volume fraction more than 0.04. Fig. 5.22 indicates the trend for transverse loss factor η_{23}. The loss factor η_{23} decreases with increase in V_I rapidly for higher value of β. Results obtained by this model shows similar trends with the FEM based evaluation with lower magnitude.

5.4.2 Optimization of fiber packing factor

The objective of the optimization of fiber packing factors α and β is to correlate the prediction of loss factors as obtained from the bridging model with that of FEM models for different fiber packing factor. FEM model has been constructed for square packing array to predict transverse and longitudinal shear loss factors and compared with the results with proposed three phase bridging model. MATLAB optimization function *Fgoalattain* is used for obtaining the fiber packing factor for different fiber volume fraction. Optimization program detail is given in Chapter 4. The longitudinal fiber packing factor (α) and the transverse fiber packing factor (β) for longitudinal shear and transverse are obtained and given in Table 5.3.

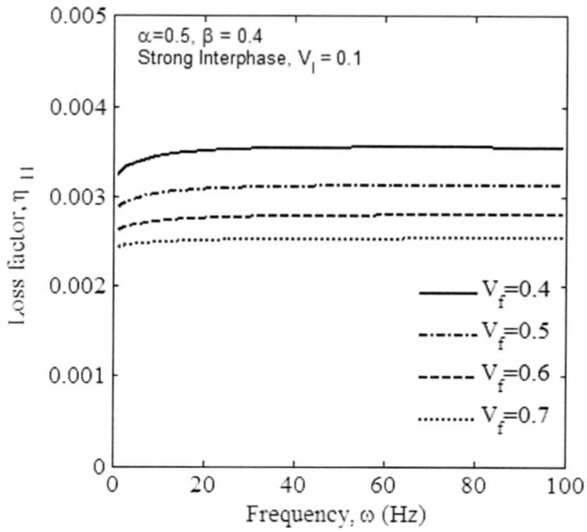

Figure 5.23 Variation of longitudinal loss factor with frequency.

5.4.3 Effect of frequency

The loss factors variation with respect to frequency by varying the fiber volume fraction and fiber packing factor with strong interphase has been discussed in this section.

Fig. 5.23 shows the variation of the longitudinal loss factor η_{11} with respect to frequency by varying the fiber volume fraction. Loss factor η_{11} increases with the increase of frequency initially for lower value of frequency and it becomes almost constant for higher value of frequency. For low magnitudes of fiber volume fraction, the effect of frequency is more in comparison to higher fiber volume fraction. For $V_f = 0.7$, loss factor η_{11} remains almost constant with frequency because at higher volume fraction, composite behavior is entirely dependent up on fiber which is elastic materials in longitudinal direction.

Fig. 5.24 shows the variation of the transverse loss factor with frequency. The loss factor η_{22} decreases rapidly initially with an increase in frequency (up to 15 Hz). Transverse loss factor is dominated by matrix property and hence its behavior is similar to matrix behavior. With the increase of fiber volume, fraction transverse loss factor decreases more at lower frequency than at higher frequency.

Transverse shear loss factor variation is shown in Fig. 5.25 with respect to frequency for strong interphase for fiber packing factor $\beta = 0.5$. Loss factor η_{23} decreases with the increase of frequency. Prediction of η_{23} is similar to

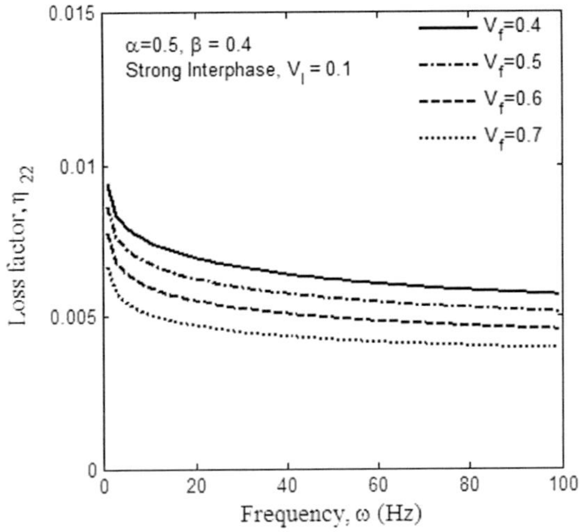

Figure 5.24 Variation of transverse loss factor with frequency.

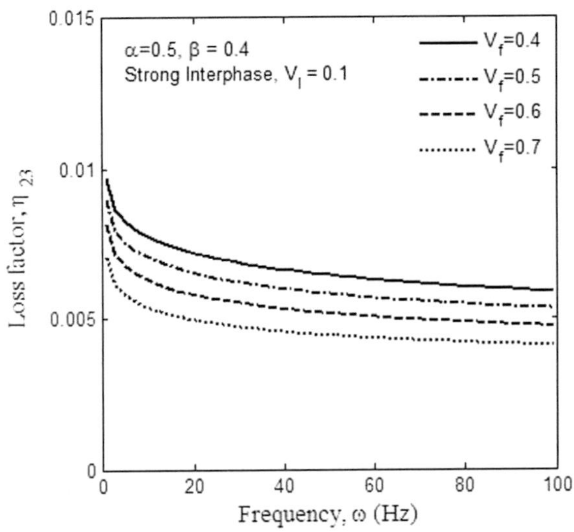

Figure 5.25 Variation of transverse shear loss factor with frequency.

transverse loss factor with higher magnitude. Decrease in transverse shear loss factor η_{23} at higher fiber fraction is more for increase in frequency (41.67% for $V_f = 0.7$ and 39% at $V_f = 0.4$). Longitudinal shear loss factor variation is shown in Fig. 5.26 with strong interphase. It decreases with the increase of frequency. Initially there is faster reduction in loss factor η_{12} in relation

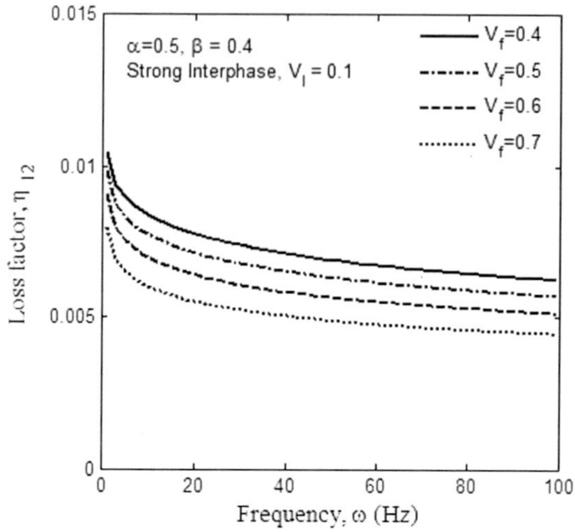

Figure 5.26 Variation of longitudinal shear loss factor with frequency.

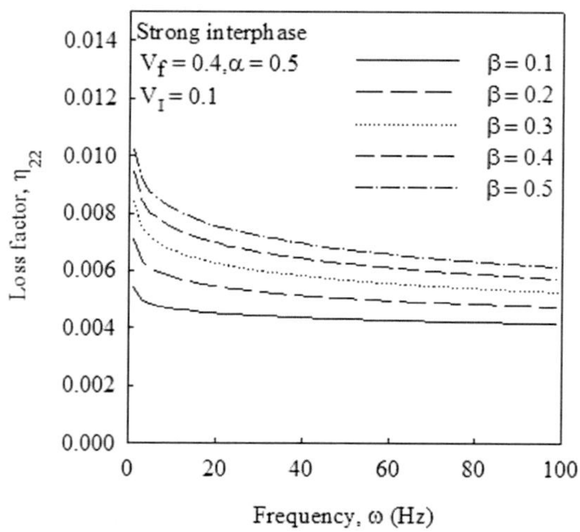

Figure 5.27 Variation of transverse loss factor with frequency for strong interphase.

to the frequency that is, 36.67% at $\omega = 5$ Hz, whereas at higher frequencies it is very small. Loss factors for lower $V_f = 0.4$ is higher as compared to the one at higher $V_f = 0.7$ corresponding to full range of frequency up to 100 Hz.

The effect of fiber packing on loss factor is depicted in Figs. 5.27–5.29 for strong interphase having 40% fiber volume fraction. Fig. 5.27 shows the variation of transverse loss factor η_{22} with frequency for increase in transverse fiber packing factor β increase from 0.1 to 0.5. Loss factor

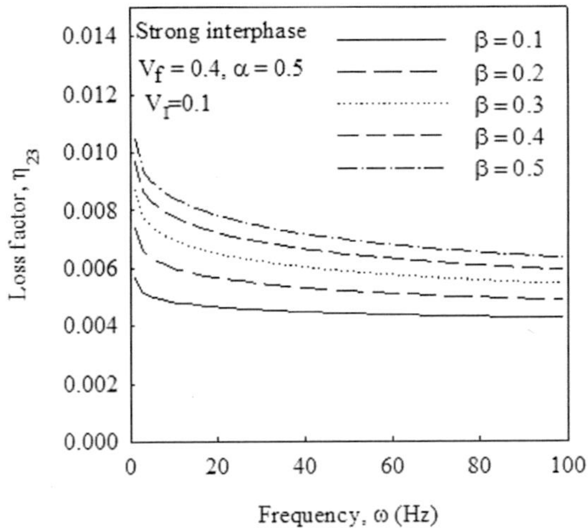

Figure 5.28 Variation of transverse shear loss factor with frequency for strong interphase.

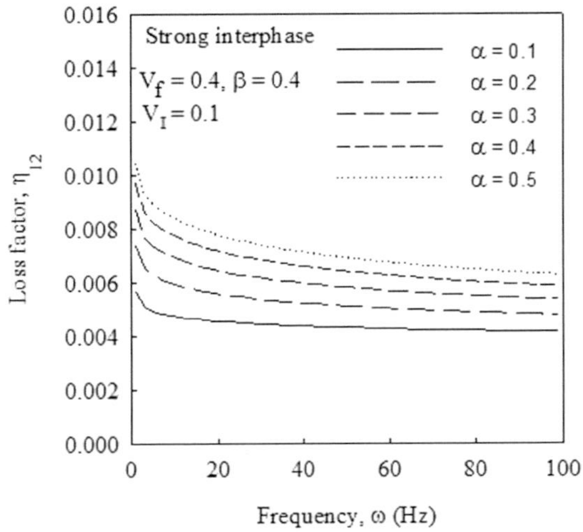

Figure 5.29 Variation of longitudinal shear loss factor with frequency for strong interphase.

η_{22} increases with increase in transverse packing factor. Transverse shear variation is shown in Fig. 5.28 and its magnitude also increases with increase in fiber packing factor, β. Increase in transverse shear loss factor is more for lower value of transverse packing factor (from $\beta = 0.1$ to 0.2 increase in loss factor is 23.76% whereas from $\beta = 0.4$ to 0.5 increase in loss factor 8.4% at frequency 11 Hz). Fig. 5.29 shows the variation of the longitudinal shear loss factor by varying the longitudinal packing factor. Loss factor η_{12} increases with an increase of fiber packing factor.

5.5 Conclusion

A mathematical model for three phase fiber-reinforced composites has been developed using the concept of a bridging model. Elastic moduli and loss factors under different loading conditions have been predicted, and a study of the effects of fiber packing factor and interphase thickness has been carried out. A comparison of results from the present model with the one existing in the literature has been made for elastic moduli and loss factors. Fiber packing plays a significant role in the prediction of elastic moduli and loss factors of the overall composite. From the comprehensive study, the following conclusions have been drawn:

(i) *Elastic modulus*

The following conclusions are made based on the comparative study of elastic moduli for strong and weak interphase by varying the fiber volume fraction:

a) Longitudinal elastic modulus has shown insensitiveness to fiber packing factor.

b) Transverse normal, transverse shear, and longitudinal shear elastic moduli decrease with the increase in fiber packing factor. For transverse packing factor equal to 1, results of transverse normal modulus and transverse shear modulus are same as CCA results. For $\beta = 0.5$ transverse normal modulus results are very near to the FEM based results reported in literature.

(ii) *Loss factors*

The following conclusions are made based on the comparative study for loss factors:

a) The longitudinal loss factor is insensitive to both longitudinal and transverse fiber packing factors.

b) The loss factors (η_{12}) increase with an increase in longitudinal fiber packing factor and loss factors (η_{22}, η_{23}) increase with an increase in transverse packing factor.

c) The longitudinal loss factor increases for strong interphase with the increase in interphase volume fraction.

d) Transverse loss factor, transverse shear loss factor, and longitudinal shear loss factor decrease marginally with increase of strong interphase volume fraction. For $\beta = 1$ transverse normal loss factor results correlate well with results reported by Vantomme.

e) For weak interphase, loss factors decrease with the increase of interphase volume fraction. Reductions in loss factors are significant for various values of fiber packing factors.

(iii) *Frequency*

The following conclusions are made on the basis of comparative study of frequency effect on composite for strong interphase:

a) The longitudinal loss factor increases with the increase of frequency.

b) The transverse loss factor, transverse shear loss factor decreases with the increase of frequency and effect of fiber packing factor is significant on loss factors.

c) The effect of frequency is more for lower fiber volume fraction.

References

[1] Z.M. Huang, Simulation of Mechanical Properties of Fibrous Composite by Bridging Micromechanical Model. Compos. Part A: Appl. Sci. Manuf. *32* (2) (2001) 143–172, doi:10.1016/S1359-835X(00)00142-1.

[2] I.C. Finegan, R.F. Gibson, Analytical Modeling of Damping at Micromechanical Level in Polymer Composites Reinforced with Coated Fibers. Compos. Sci. Technol. *60* (7) (2000) 1077–1084, doi:10.1016/S0266-3538(00)00003-8.

[3] Y. Benveniste, A new approach to the application of Mori-Tanaka's theory in composite materials, Mechanics of Materials, Vol. 6, (1987) pp. 147–57.

[4] G. Wacker, A.K. Bledzki, A. Chate, Effect of Interphase on the Transverse Young's Modulus of Glass/Epoxy Composites. Compos. Part A: Appl. Sci. Manuf. *29* (5–6) (1998) 619–626, doi:10.1016/S1359-835X(97)00116-4.

[5] J. Vantomme, Parametric Study of Material Damping in Fiber Reinforced Plastics. Composites *26* (2) (1995) 147–153, doi:10.1016/0010-4361(95)90415-V.

[6] R. Chandra, S.P. Singh, K. Gupta, Micromechanical Damping Models for Fiber Reinforced Composites: A Comparative Study. Composite: Part A 33 (2002) 789–796.

CHAPTER 6

Modeling of nonlinear damping

In this chapter, a nonlinear model has been proposed for the prediction of damping of viscoelastic materials. Hysteretic loop has been obtained for evaluation of damping of the fiber-reinforced composite materials using proposed nonlinear modified four parameter model for viscoelastic materials. This model is used for prediction of loss factors for different fiber volume fraction of the fiber-reinforced composite.

6.1 Introduction

Damping capacity is a measure of a material's ability to dissipate energy during mechanical vibration under cyclic loading [1]. The damping of conventional materials used in conventional ways does not generally provide sufficient energy dissipation for many types of engineering structures. Composite materials are used for improved tailoring of properties. Polymers generally display linear behavior in the region of low and intermediate stress. Hysteretic loops remain elliptical. However, significant nonlinearity had been shown by most materials at high strain. Many authors [2,3] have evaluated nonlinear damping by different methods. The nonlinear behavior of polymeric fiber composite materials has been studied theoretically in terms of a model developed for elastic plastic materials, and generally valid for elastic–plastic response. This theory is focused on the current material state and connects material anisotropic response with identifiable directions in the present material state [4]. Pooler and Smith [5] studied the effect of time and temperature on a commercial wood plastic composite by means of series of creep and recovery tests. A prony series was used to describe the material time dependant compliance where times are shifted with stress and temperature to describe nonlinear response and temperature dependence.

Here, a nonlinear model has been proposed for the prediction of damping of viscoelastic materials. Hysteretic loop has been obtained for the evaluation of damping of the materials. Results are obtained for nonlinear four parameter model for viscoelastic materials. This model is used for prediction of

Damping in Fiber Reinforced Composite Materials.
DOI: https://doi.org/10.1016/B978-0-323-91186-3.00002-2

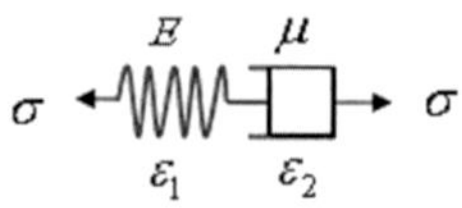

Figure 6.1 Maxwell model.

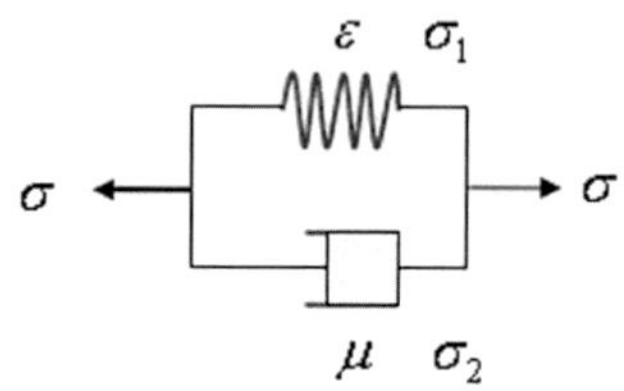

Figure 6.2 Voigt model.

damping of the fiber-reinforced composite. Further the variation of strain with respect to time has been evaluated.

6.2 Mathematical model

Mathematical model development for rheological behavior of solid is to permit the realistic results to obtain from mathematical analysis of complicated structures under various conditions, such as sinusoidal, random, and transient loading. In 1784, Coulumb recognized that the mechanism of damping operative at low stresses may be different than those at high stresses. Even today major emphasis is placed on linear models of damping because of its accuracy at low stress regime and computationally more economical than nonlinear ones. The linear model for viscoelastic material is Maxwell model, which consist of a spring and dashpot in mechanical series as shown in Fig. 6.1. Further Kelvin–Voigt model which comprise of a spring in parallel with dashpot as shown in Fig. 6.2. After two parameter, three parameter and four parameters model have been developed that predict more accurately the real behavior of viscoelastic solid at small amplitude of strain or stress polymers, elastomers and metals generally display linear rate dependent damping that can be simulated by four parameter viscoelastic model. However, significant nonlinearity is displayed by materials at high strain. To simulate the nonlinear behavior of materials, four parameter must include the nonlinear component. The differential equation derived for nonlinear modeling is discussed below:

Modified form of four parameter viscoelastic model which consist of two spring and two dashpot as shown in Fig. 6.3. This model is formed by combining the Maxwell model (consist of spring and dashpot in series as shown in Fig. 6.1) and Voigt model (where spring and dashpot are connected

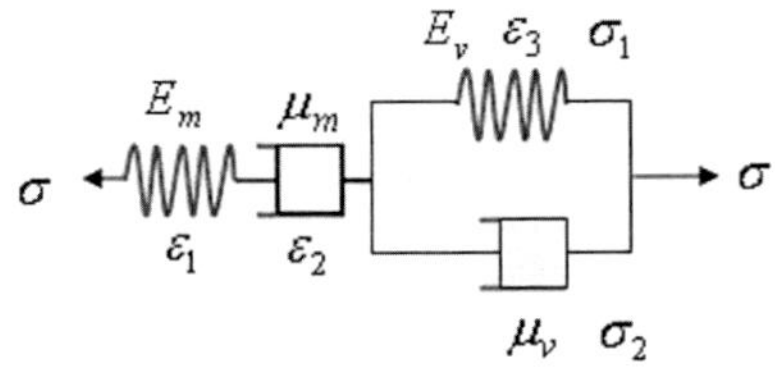

Figure 6.3 Four parameter viscoelastic model.

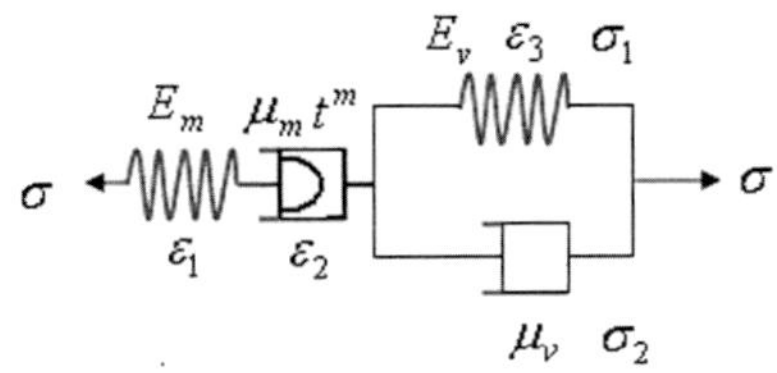

Figure 6.4 Four parameter nonlinear viscoelastic model.

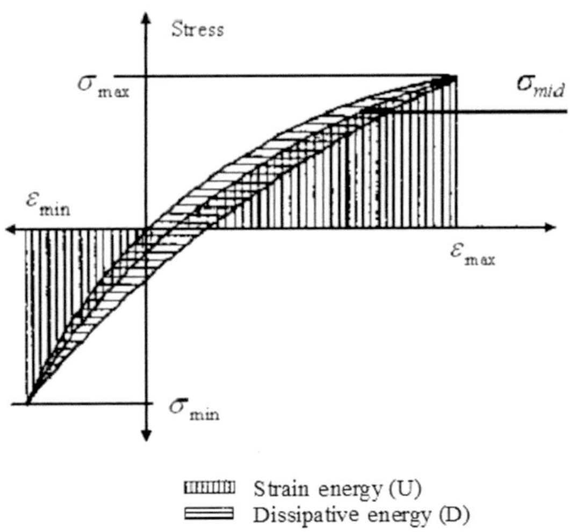

Figure 6.5 Hysteresis loop for damping measurement.

in parallel shown in Fig. 6.2) model in series.

$$\sigma = E\varepsilon_1,\ \sigma = \mu\varepsilon_2 \tag{6.1}$$

Where σ is stress, is ε strain and ε is strain rate. E is elastic modulus and μ is the viscosity coefficient.

$$\varepsilon = \varepsilon_1 + \varepsilon_2 \tag{6.2}$$

$$\frac{\sigma}{E} + \frac{\sigma}{\mu} = \varepsilon_1 + \varepsilon_2 = \varepsilon \tag{6.3}$$

Eq. (6.3) is written in standard format as below

$$\sigma + p_1\sigma = q_1\varepsilon \tag{6.4}$$

Further equation for stress strain relation for Voigt model as shown in Fig. 6.2 is derived as given below:

$$\sigma_1 = E\varepsilon, \quad \sigma_2 = \mu\varepsilon \tag{6.5}$$

$$\sigma = \sigma_1 + \sigma_2 = (E\varepsilon + \mu\varepsilon) \tag{6.6}$$

This equation is written in standard form as in Eq. (6.7)

$$\sigma = q_0\varepsilon + q_1\varepsilon \tag{6.7}$$

Four parameter model are formed by combing these two model in series, as shown in Fig. 6.3 and differential equation is derived as below:

$$\varepsilon = \varepsilon_1 + \varepsilon_2 + \varepsilon_3 \tag{6.8}$$

$$\sigma = E_m\varepsilon_1, \quad \sigma = \mu_m\varepsilon_2 \tag{6.9}$$

$$\sigma = \sigma_1 + \sigma_2 = (E_v\varepsilon_3 + \mu_v\varepsilon_3) \tag{6.10}$$

Laplace transformation is applied on both side of the Eq. (6.9) and Eq. (6.10)

$$\sigma = \mu_m s\varepsilon_2 \tag{6.11}$$

$$\sigma = (E_v + \mu_v s)\varepsilon_3 \tag{6.12}$$

Multiplying each of these equations with suitable constant and adding we get

$$\begin{aligned}\sigma E_m(\mu_m s + E_v + \mu_m s) + \sigma\left(E_v\mu_m s + \mu_v\mu_m s^2\right) \\ = \left(E_v\mu_m s + \mu_v\mu_m s^2\right)E_m\varepsilon\end{aligned} \tag{6.13}$$

$$\sigma + p_1\sigma + p_2\sigma = q_1\varepsilon + q_2\varepsilon \tag{6.14}$$

where $p_1 = \frac{\mu_m}{E_m} + \frac{\mu_v}{E_v} + \frac{\mu_m}{E_v}$, $p_2 = \frac{\mu_m\mu_v}{E_m E_v}$, $q_1 = \mu_m$, $q_2 = \frac{\mu_m\mu_v}{E_v}$

E_m, E_v are Maxwell and Voigt elastic modulus and μ_m, μ_v are Maxwell and Voigt viscosity of assumed four parameter model for viscoelastic material.

For incorporating the nonlinear phenomena, one dashpot is replaced with nonlinear dashpot which considers the ageing effect on materials as viscosity coefficient of this dashpot is time dependent and is given by Eq. (6.15) [1]

$$\sigma = \mu\, t^{m}(\varepsilon) \tag{6.15}$$

Where m is the aging exponent. For m is negative material display time hardening. If m is positive, materials display time softening viscosity [1]. In Eq. (6.14) q_1 is replaced with $q_1 t^{m}$. Hence, equation for nonlinear four parameter model is given as in Eq. (6.16).

$$\sigma + p_1\sigma + p_2\sigma = q_1 t^{m}\,\varepsilon + q_2\varepsilon \tag{6.16}$$

Solving this Eq. (6.16), stress and strain for cyclic loading have been obtained with the help of Eq. (4.53) and (4.54). Using this model, polymer composite is then modeled through bridging model (Haung [6]). This model is analysed under cyclic loading and the results are obtained and plotted in form of hysteresis loop.

6.2.1 Hysteresis loops for vibrating systems

In study of vibration, hysteresis loop is very often used as measure of energy dissipation. A hysteresis loop is obtained by recording the magnitude of a force verses the displacement brought about by its action. The area enclosed by the loop is proportional to the amount of energy dissipated in the system within one period as given by Eq. (6.19).

$$\text{Dynamic Modulus} = \frac{\sigma_{\max} - \sigma_{\min}}{\varepsilon_{\max} - \varepsilon_{\min}} \tag{6.17}$$

$$\text{Strain energy} = \int_{\varepsilon=0}^{\varepsilon_a} \sigma\, d\varepsilon \tag{6.18}$$

$$\text{Dissipative Energy} = \oint \sigma\, d\varepsilon \tag{6.19}$$

Solving nonlinear Eq. (6.16), the stress and strain is obtained. Loss factor has been obtained by hysteretic loop plotted by using the Eq. (6.20) [1].

$$\eta = \frac{D}{2\pi U} = \left[\int_{0}^{2\pi/w} \sigma\,(d\varepsilon/dt)dt\right]\left[2\pi \int_{\varepsilon=0}^{\varepsilon} \sigma_{mid}\, d\varepsilon\right]^{-1} \tag{6.20}$$

Where σ_{mid} = mid stress

Table 6.1 Material properties used for simulation.

E_f (GPa)	E_m (GPa)	E_v (GPa)	μ_m (GPa h)	μ_v (GPa h)
72.4	3	3	21,000	3

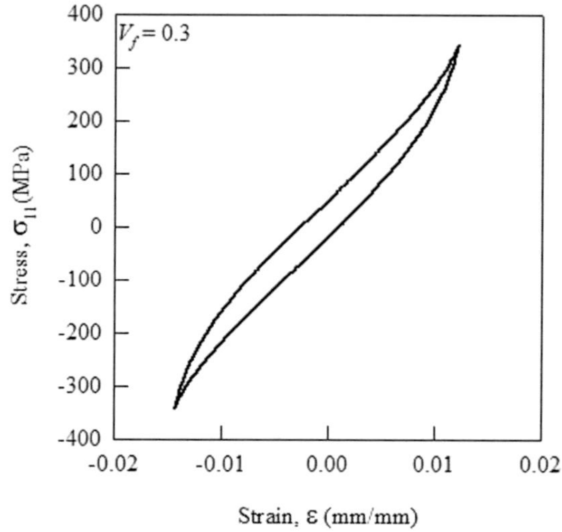

Figure 6.6 Variation of longitudinal stress with respect to strain for $V_f = 0.3$.

6.3 Results and discussions

Using the material properties given in Table 6.1, stresses and strains are evaluated for different fiber volume fractions from $V_f = 0.3$ to $V_f = 0.7$. Strain varies linearly for very low values of stress and it takes sharp variation at higher stresses.

Fig. 6.6 shows the variation of longitudinal stress with respect to longitudinal strain for fiber volume fraction $V_f = 0.3$. Stress is completely reversible from maximum to minimum. Stress vary linearly for small strain and its changes to nonlinear form for higher strain. Figs. 6.7–6.10 show the variation of longitudinal stress with respect to strain for fiber volume fraction $V_f = 0.4$, $V_f = 0.5$, $V_f = 0.6$, $V_f = 0.7$, $V_f = 0.8$ respectively. With increase of fiber volume fraction, area under hysteresis loop is getting smaller and smaller. This indicates that damping of the fiber-reinforced composite materials is reducing with increase of fiber volume fraction. Figs. 6.12–6.15 demonstrate the variation of transverse stress with respect to strain in transverse direction for fiber volume fraction $V_f = 0.3$, 0.4, 0.5, and 0.6 respectively for GFRE composite for fiber orientation $\theta = 90^0$. Slope of

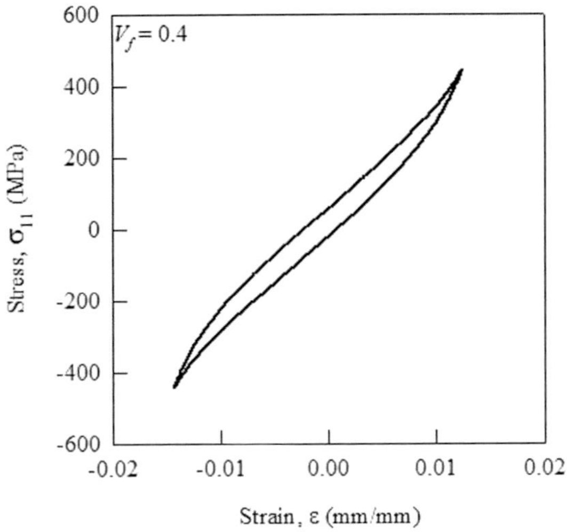

Figure 6.7 Variation of longitudinal stress with respect to strain for $V_f = 0.4$.

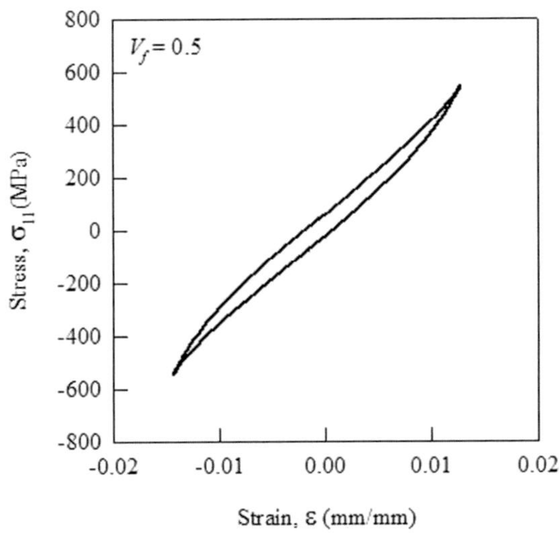

Figure 6.8 Variation of longitudinal stress with respect to strain for $V_f = 0.5$.

the stress variation is less and loop area is bigger than that of longitudinal loading.

Fig. 6.16 shows the variation of loss factors (η_{11}, η_{22}) with fiber volume fraction. Loss factor is decreasing with the increase of fiber volume fraction. Table 6.2 shows the variation of longitudinal and transverse loss factors with

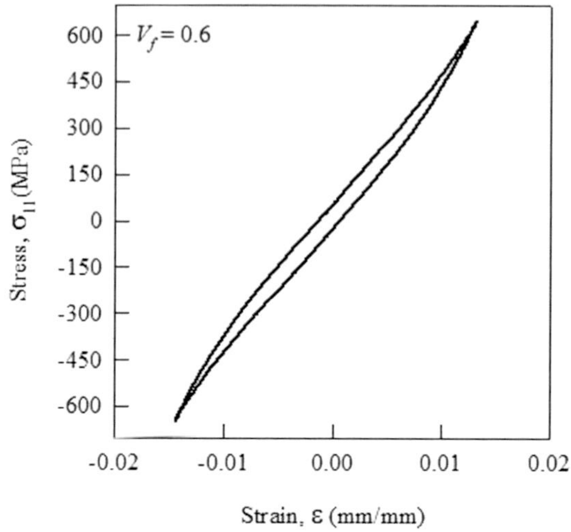

Figure 6.9 Variation of longitudinal stress with respect to strain for $V_f = 0.6$.

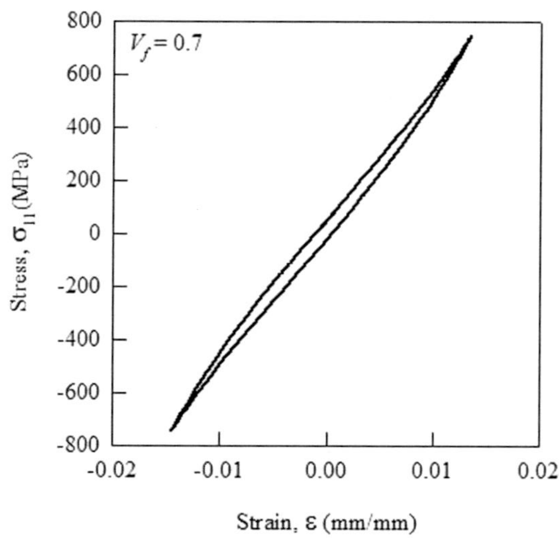

Figure 6.10 Variation of longitudinal stress with respect to strain for $V_f = 0.7$.

respect to fiber volume fraction. It can be seen that the percentage decreases in loss factors η_{11} is of the order of 64% as the volume fraction increases from 0.3–0.7. Similarly reduction in loss factor η_{22} is of the order of 17% for the change in fiber volume fraction from 0.3 to 0.7. Fig. 6.17 shows the variation of stress with respect to strain for constant stress by increasing the

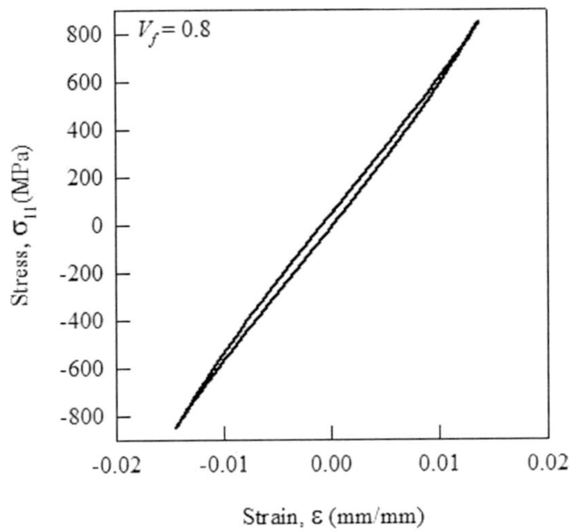

Figure 6.11 Variation of longitudinal stress with respect to strain for $V_f = 0.8$.

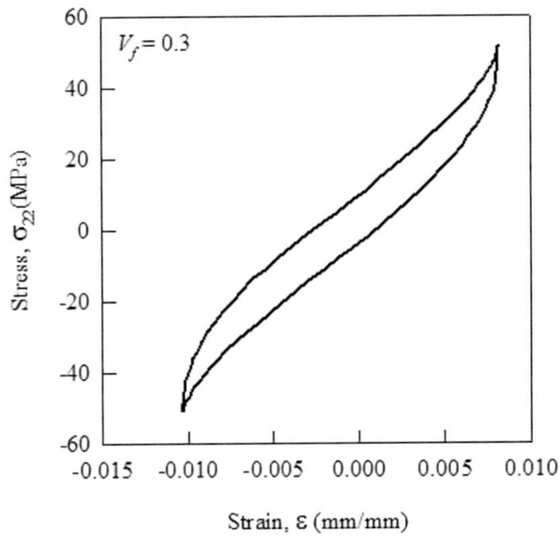

Figure 6.12 Variation of transverse stress with respect to strain for $V_f = 0.3$.

strain in each cycle. Area of the loop is increasing with the increase of strain and hence damping is increasing with increase of strain. Material is getting softer with time [7]. Fig. 6.18 depicts the variation of longitudinal loss factor with respect to strain. Longitudinal loss factor increases rapidly initially with strain and then its increase in loss factor is steady with the strain. It can

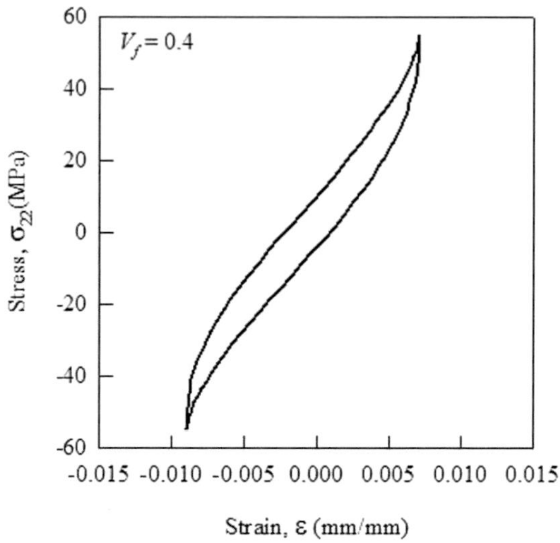

Figure 6.13 Variation of transverse stress with respect to strain for $V_f = 0.4$.

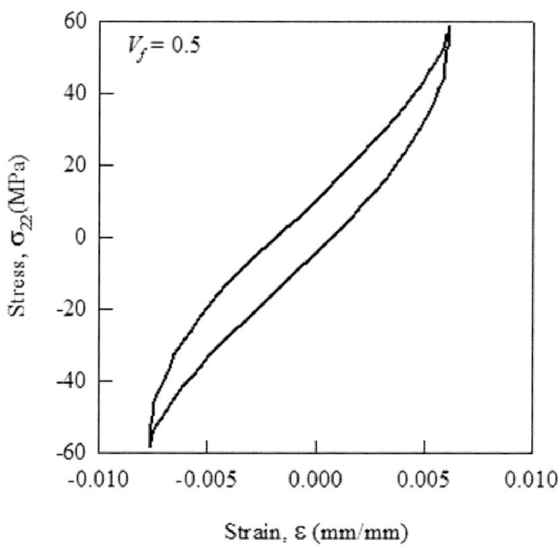

Figure 6.14 Variation of transverse stress with respect to strain for $V_f = 0.5$.

be observed from the results that the variation of longitudinal loss factor is of the order of 70% as strain varies from 0.01 to 0.017 whereas amount of increase in damping is of the order of 27 % for the increase in strains from 0.02 to 0.039. This indicates that the nature of nonlinearity of damping is quite high in case of longitudinal loss factor.

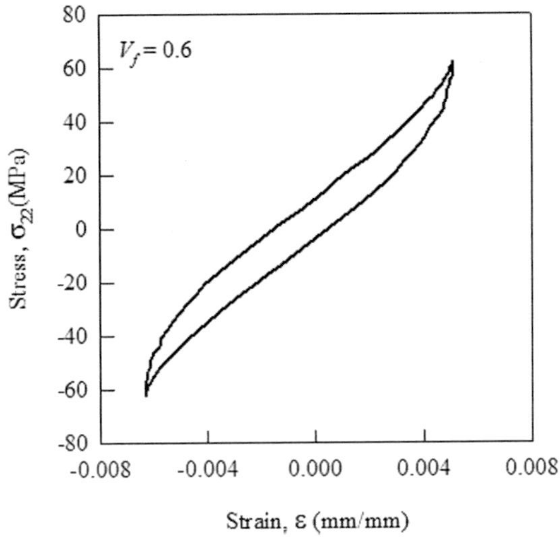

Figure 6.15 Variation of transverse stress with respect to strain for $V_f = 0.6$.

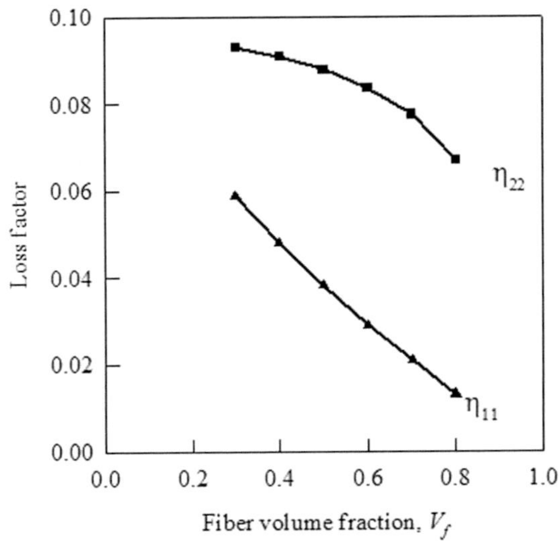

Figure 6.16 Variation of loss factor with respect to fiber volume fraction.

Table 6.2 Loss factor variation with fiber volume fraction.

Fiber volume fraction, V_f	Loss factor, η_{11}	Loss factor, η_{22}
0.3000	0.0587	0.0931
0.4000	0.0480	0.0909
0.5000	0.0382	0.0879
0.6000	0.0289	0.0837
0.7000	0.0210	0.0776
0.8000	0.0131	0.0671

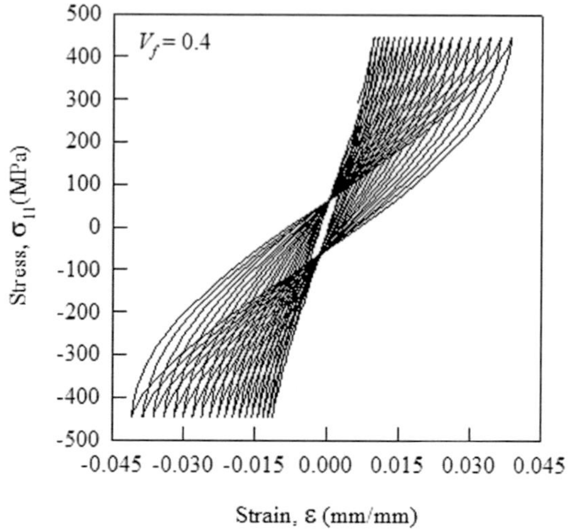

Figure 6.17 Variation of stress with respect to strain for different cycle.

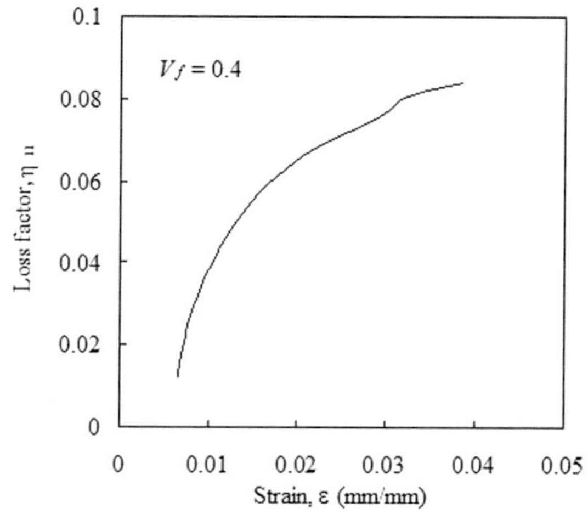

Figure 6.18 Variation of longitudinal loss factor with respect to strain.

6.4 Conclusion

A nonlinear four parameter model has been developed for viscoelastic materials. Nonlinear four parameter viscoelastic model has been applied to Bridging model in order to evaluate longitudinal and transverse loss factor for unidirectional glass fiber-reinforced epoxy composite using the hysteresis loop method. The following conclusions are drawn based on results obtained:

1. Longitudinal loss factor decreases with the increase of fiber volume fraction whereas it increases with increase of strain.
2. Transverse loss factor decreases with the increase of fiber volume fraction.
3. Transverse loss factor increases with increase of strain.
4. Longitudinal and transverse are quite higher for higher strains and show nonlinearity.

References

[1] J.B. Lazan, Damping of Materials and Members in Structural Mechanics, Pergaman Press, Oxford, New York, 1968.
[2] A.D. Drozdov, A constitutive model for nonlinear viscoelastic media, Int. J. Solids Struct. 35 (18) (1998) 2315–2347.
[3] A.M. Horr, L.C. Schmidt, Modelling of nonlinear damping characteristics of a viscoelastic structural damper, Eng. Struct. 18 (2) (1996) 154–161.
[4] G. Spathis, E. Kontou, Non-linear viscoplastic behavior of fiber reinforced polymer composites, Compos. Sci. Technol. 64 (15) (2004) 2333–2340.
[5] D.J. Pooler, L.V. Smith, Nonlinear viscoelastic response of wood plastic composite including temperature effect, J. Thermoplast. Compos. Mater. 17 (2004) 427–444.
[6] Z.M. Huang, Simulation of mechanical properties of fibrous composite by bridging micromechanical model, Composites: Part A 32 (2001) 143–172.
[7] Z. Osinski, Damping of Vibrations, A.A. Balkema, Rotterdam, 1998.

CHAPTER 7

Applications of damping and principles of vibration control

During the last few decades, the application of fiber-reinforced polymer (FRP) composites in various engineering fields, viz. aeronautics, automotive, sports, renewable energy, and construction has increased dramatically. This is mainly due to the high specific properties of the FRPs. The FRPs are subjected to cyclic and impact loading during their operation which results in vibrations leading to the shortening of their lifespan. The damping performance of FRPs is therefore a critical parameter that has gained significant attention of the researchers worldwide. Some of the main engineering applications of damping in real life have been discussed below. The readers will be able to infer the importance of damping and principles of vibration control in FRPs after reading this chapter.

7.1 Applications of damping

Hadiji et al. [1] evaluated the dynamic properties of nonwoven flax, hemp, kenaf, and glass fiber-reinforced polypropylene (PP) composites. Also, the influence of some parameters, such as the type of reinforcement, the fiber/matrix weight ratio, the fiber orientation and the porosity content, on the damping behavior of these nonwoven composites was investigated. A free flexural vibrations analysis was conducted to experimentally identify their natural frequencies and their associated loss factors. The obtained results show that the nonwoven composites reinforced by natural fibers present higher loss factors compared with those of the glass-PP composites. These interesting damping properties make these nonwoven composites very attractive for *automotive applications* where the dissipation of vibrations is highly requested.

Because of their lightweight and high strength, carbon fiber-reinforced plastics (CFRPs) are used with more than 50 wt. % as main structural parts, for example, *fuselage and wing sections, in Boeing 787s*. Indeed, they have long been used in high-speed trains, for example, nose and hood. As such,

Damping in Fiber Reinforced Composite Materials.
DOI: https://doi.org/10.1016/B978-0-323-91186-3.00006-X

they are often subjected to cyclic, periodic and impact loads during service, thus generating different levels of structural vibration and noise. To alleviate these problems, high damping materials with good capacities to dissipate the mechanical energy due to the vibration-induced loading by conversion to other forms of energy (normally heat) are required. Unfortunately, CFRPs have high stiffness/high strength but low damping capacity due to the poor viscoelastic characteristics of the carbon fibers. Hence, enhancement of damping properties of CFRPs to meet the stringent use requirement of engineering structures when subjected to vibrating loads is of critical importance. Li et al. [2] fabricated polysulfone (PSF)/cellulose nanocrystal (CNC) nanofibrous membranes by electrospinning as hybrid interleaves in carbon fiber/epoxy (CF/EP) composites. The damping properties of the interleaved CF/EP composites were measured by the cantilever percussion free-decay method and dynamic mechanical analysis (DMA). The strength and modulus values of the interleaved composites were obtained by tensile and flexural tests. The obtained results showed that after interleaving by the 10%PSF/0.5%CNC membranes, the damping properties of CF/EP composites were enhanced with little loss of mechanical properties. The damping mechanisms were examined and analyzed with the aid of scanning electron microscopy (SEM). It was concluded that the enhanced damping properties were due to the different energy dissipation mechanisms associated with the PSF nanoparticles generated by reaction induced phase separation and the incorporated CNCs in the EP matrix.

The service life of *concrete structures* is often threatened by unfavorable dynamic loads. Optimizing the material damping can improve the vibration damping of the structure. Khayat et al. [3] developed a polyvinyl alcohol fiber-reinforced high-volume fly ash (PVA-HVFA) cementitious composite with high damping characteristics and moderate mechanical properties. The coupling effects of frequency and temperature on the damping were further clarified. Experiments combined with MIP and SEM tests were performed to analyze the pore structure distribution, fiber–matrix morphology, and enhanced damping mechanism. The results showed that the large amount of FA adversely affected strengths of the PVA-HVFA but significantly improved the damping and energy dissipation capacity. Meanwhile, the PVA fibers enhanced both strengths and damping characteristics. Vibration damping and energy dissipation were hypothesized to be affected by the matrix, fibers, and fiber–matrix interface. PVA-HVFA is expected to be applicable to beam–column plastic hinges and other vibrational damping fields.

The viscoelastic damping material is inserted into structural surface to reduce vibration and noise in aerospace, automobile and residential structures. The constrained damping layer undergoes shear deformation when structure is subjected to bending vibration. The structure dissipates energy through shear deformation to increase its damping. Thus, the vibration can be controlled to prolong fatigue life of structure while maintaining its functional integrity. Although a large number of scholars have done a lot of research on the damping property of sandwich composites, there are few researches on how to enhance their interfacial bonding property. In addition, the damping properties of sandwich composites studied in the past are improved at the cost of sacrificing their mechanical properties. Moreover, traditional co-cured damping composite shows serious defects, such as aging of damping layer and weak interface adhesion between sandwich and composite layer. Therefore, in order to improve damping performance and mechanical property of traditional sandwich composites and prolong their service life, the novel co-cured epoxy resin matrix carbon fiber-reinforced damping composites (CRFRDC) were investigated by Liang et al. [4]. The novel CRFRDC were prepared via co-curing process. In addition, it is necessary to convert damping material into a solution during preparation of CRFRDC so that damping solution can permeate into the prepreg. Compared with the traditional co-curing damping composite structure, active functional groups in rubber and resin matrix can react chemically under the action of high temperature, high pressure, and catalyst. The rubber sandwich material and epoxy resin can be combined by physical fusion or chemical cross-linking. Therefore, CRFRDC not only maintain the mechanical performance of traditional composites, but also dramatically enhance their interfacial bonding properties. Meanwhile, CRFRDC show large specific modulus, aging resistance and superior damping performance compared with traditional sandwich composites. It exhibits a broad application prospect in the field of *high-grade automobile and engineering vibration reduction*. The co-cured damping composite structure in the work by Liang et al. [4] shows the characteristics of large damping and high stiffness, and solves the issue of contradiction between stiffness and damping. It has a broad application potential in the field of *aerospace and high-speed railway*.

As a basic unit, the FRP composite plate has been widely used in *nautical, aviation, aerospace, and other engineering fields*. As the FRP composite plate is frequently working in harsh loading conditions, it is prone to generating excessive vibration phenomenon and dynamic stress, which leads to the delamination and the fatigue failure issues of composite structures.

Therefore, it is necessary to study the corresponding vibration suppression techniques. There are many approaches for vibration suppression of composite structures, such as shifting the structural quality and rigidity, isolating the transmission of vibration energy, and dynamic absorption techniques. However, for the FRP composite plate structure, those above approaches are still far from satisfaction in achieving desirable results due to the limitations of volume, mass, and other factors. Damping coating technology is a new and effective way for vibration suppression of various composite structures, which has attracted the attention of various researchers. It has been reported that the functional graded porous graphene coating has better excellent mechanical property and thermal resistance. Therefore, the application of this coating to the surface of the FRP composite plate will be able to strengthen antivibration capability and then improve the structural service life. Due to the excellent energy dissipation effect, the viscoelastic foundation is discovered to be another effective way to enhance the vibration suppression capacity of the composite structure. Since the boundary condition has a dominating effect on the modal shape, stiffness, and damping capacity of the composite structure, it is also possible to reduce the structural vibration response by altering the boundary conditions. Zhao et al. [5] investigated the vibration suppression effects of the viscoelastic foundation, viscoelastic damping boundary condition and porous graphene platelet coating (PGPC) on the FRP composite plate. A dynamic model was built by combining the first-order shear deformation theory, the multi-segment partition technique, the virtual spring technology, the Rayleigh-Ritz approach, and the mode strain energy method. Afterwards, the structural natural frequencies, modal shapes and modal damping ratios were calculated, as well as the verification of the convergence and the correctness of the constructed model. Finally, the vibration suppression effect of the FRP-PGPC composite plate with different parameters, such as geometrical and material parameters, foundation and boundary condition types, were discussed in detail.

As a key structural component, the drive shaft plays an important role in transmitting torque and output power, which is widely used in aerospace, shipbuilding, automobile, wind power, and other fields. Compared with the traditional metal material drive shaft, the FRP drive shafts have a higher specific strength, specific stiffness, fatigue resistance, design flexibility, and high natural frequency. Considering the high-speed rotating parts of the drive shafts due to the unbalanced mass, the error of the transmission system and the fluctuation factors of the working load often cause the vibration of the structural components and noise generation. When the vibration frequency the natural frequency of the material, it will also cause resonance damage to the material and fatigue failure of drive shafts in the service environment. Therefore, vibration performance is one of the key factors to be considered in the design and manufacture of high-speed transmission shafts. Sun et al. [6] used the finite element analysis (FEA), pulse vibration excitation technique (PVET) and formula computing to study the effect of the vibration characteristics of shafts. Their study revealed that increasing the modulus of the fiber is one of the effective methods to increase the natural frequency of the FRP tube and the damping ratio of the carbon FRP tubes were slightly smaller compared with basalt and glass fiber. The FEA of FRP tube results were not distinct from the measured and calculated results so that these FEA results were relatively reliable. The damping of the tube with metal flanges at both ends were weaken compared the FRP tube. This work will be of great significance to the optimization design of the vibration performance of FRP shafts.

Carbon fiber-reinforced epoxy composite (C-epoxy) is a subclass of CFRPs which are excessively being used nowadays in the area of aerospace, automotive and underwater applications to replace conventional metallic materials because of their high strength to weight ratio and rigidity. Though C-epoxy composites have high specific strength they are relatively brittle in nature due to high crosslink density with poor resistance to the initiation and growth of cracks. Therefore efforts were made for toughening of these brittle C-epoxy composites through various methods. Among them the most prominent method is the addition of phase separated rubber particles to the matrix system. Various liquid rubber materials, such as butadiene acrylonitrile co-polymers containing carboxyl (CTBN), amine terminated butadiene (ATBN) and epoxy terminated butadiene (ETBN) reactive end groups are generally added. Rao et al. [7] used epoxy resin modified with epoxy terminated butadiene acrylonitrile liquid rubber to make carbon fiber-reinforced epoxy (C-epoxy) composites with an aim to enhance the impact, compression and damping properties. ETBN added C-epoxy composites were further modified using amino functionalized multi-walled carbon nanotubes (A-MWCNTs) at different weight ratios namely 0.5 wt. %, 1.0 wt. %, and 1.5 wt. % and impact, compression and damping properties were predicted. Results indicate that modification of C-epoxy with ETBN enhances the properties considerably. A-MWCNTs addition to ETBN modified C-epoxy (A-MWCNTs-C-Epoxy-ETBN) further enhances the impact strength (26%), compressive strength (17%) at 1.0 wt. % and damping ratio (87%) at 0.5 wt. % of A-MWCNTs. The material developed is suitable for the underwater applications, such as shells and other mechanical systems. The underwater mechanical systems, such as weapon systems experiences shock loads when entering into the water during launching. So the materials should have high impact strength to overcome these shock loads. Damping is also critical for the materials used in underwater applications, as its improvement will reduce the noise levels and enhances stealth properties.

Biodegradable natural fiber composites are increasingly developed for lightweight structural applications in automobile and construction industries with significant environmental benefit. Wang et al. [8] explored the sound and vibration properties of natural fiber composites that inevitably suffer menace of vibrations during their operations as engineering structures. Natural fibers (bamboo, cotton, flax) and polylactic acid (PLA) fibers were blended, carded, laid up and finally needle-punched to nonwoven mats before being consolidated at elevated temperature under compression. Vibration damping and acoustic measurements were conducted on these natural fiber composites and their behavior was compared to that of a commercial panel, made of PP matrix reinforced with hemp and kenaf fibers. From both the coincidence frequency analysis and wave number amplitude analysis, the cotton/bamboo/PLA composite laminate showed the best acoustic performance (coincidence frequency of 2448 Hz) that is related to the fineness of cotton fibers, bending stiffness, natural frequency and density of the composite panel. The results suggest that the bamboo/cotton hybrid composites can be a viable candidate for engineering applications that require good bending properties, high sound absorption and vibration damping properties.

Rubber composites provide significant properties, such as high flexibility, lightweight, high tensile strength, and substantial damping. Rubbers are often used to control vibrations and impacts in machinery and electronic devices due to their impact absorption characteristics. Rubbers are widely used in different industries, including oil and gas, automotive, shipping and maritime, railways and aviation, road and bridge construction and impact energy absorption. Rubbers are used along with metal or composite layers as a solution to improve impact energy absorption in structures. Liaghat et al. [9] investigated the effect of incorporating different fibers in an elastomeric media on its mechanical and energy absorption characteristics. To this end, newly developed compounded natural rubber was employed as the matrix and three chopped fibers including glass, carbon and Kevlar were used as the reinforcement at different weight loadings. Dynamic tensile tests revealed that while carbon and glass fibers had similar influence in reducing both strength and failure strain of the composites, Kevlar fibers contributed positively to increasing the ultimate strength. In general, energy absorption capacity of fiber-reinforced rubber was observed to be less than that of neat specimens. Based on the dynamic indentation tests, incorporating fibers within the elastomeric matrix could be beneficial in small deformations owing to the more energy absorption capacity of fiber composites at the beginning of the indentation process. The optimum configuration was concluded to be 5 wt.% of carbon fibers in which the deflection-to-failure is improved whilst the energy absorption and ultimate strength remain intact.

Fiber-reinforced composites offer a strong and lightweight alternative to traditional building materials and are increasingly being used across a broad spectrum of engineering applications, including aerospace structures, sports equipment, and wind turbines. Composite materials used in such applications often suffer from vibrations caused by dynamic loads acting on the system, resulting in reduced efficiency and comfort, undesirable noise, and a shortened lifespan of the structure. Previous research has shown that various types of nanoscale fillers, such as carbon nanofibers, ceramic nanoparticles, and carbon nanotubes (CNTs) can be incorporated into the composite matrix to enhanced the electrical, thermal, and mechanical properties of the material, including vibration-damping augmentation. In particular, the addition of CNTs to composite matrices has been found in several studies to provide significant damping enhancement to the material, allowing for the fabrication of composite structures with increased resistance to vibrations. Pitchumani and DeValve [10] experimentally investigated the damping effects of CNTs embedded in the matrix of fiber-reinforced composite materials. Several different aspect ratios, types, and weight fractions of CNTs were considered, and an analysis of the CNT dispersion in the composite matrix was presented. The composite materials were analyzed using dynamic mechanical analysis and various modal analysis techniques to determine the damping characteristics of the composite as a function of strain, fiber volume fraction, and nanotube type and weight percentage loading. Experiments were conducted using cantilevered beams in both a stationary and rotating frame in order to explore the effects of rotation on the damping behavior of the composite material. The results showed that the addition of two weight percent of CNTs to the matrix of carbon fiber-reinforced composites can increase the damping in a stationary composite beam by more than 130% and by more than 150% in a composite beam rotating at 500 RPM.

Advanced materials, such as plastics and composites, have been invented and adopted for bespoke products in railway industry. For instance, they have been employed as railway track sleepers, switches and crossings' bearers, bridge transom girders, and so on. Railway track sleepers and bearers are essential for operations that enhance safety and reliability of railway turnouts. The degradation mechanism of railway sleepers/bearers is attributable largely to the type of materials used. The utilization of composite material as turnout bearers, located in railway switches and crossings, has resulted in many questions whether they can experience aggressive environments. In reality, the dynamic properties of bearers play a key role in the serviceability performance of the switches and crossings. Exposed to high-intensity impact loading condition, switches and crossings withstand excessive vibrations, which can disable craw lock or signaling gear systems. Even though such effects are well known, the influence of environmental variance on the dynamic properties of the bearers has yet been determined thoroughly. Kaewunruen et al. [11] investigated the influence of wet and dry conditions for the practical guideline and design solutions using composite materials for railway turnout systems. Fiber-reinforced foamed urethane (FFU), was emphasized since FFU has been installed in several railway switches and crossings globally. The dynamic properties of the FFU materials were identified using the instrumented impact hammer testing method. The dynamic damping property of FFU was calculated by exponential curve fittings, and logarithmic decrements. The results showed that FFU on average absorbs less than 5% of water. Interestingly, the water absorption led to the increment in dynamic moduli in both fiber-grain directions up to 16%.

Since 1950s, FRP matrix composites have been used in marine environment. Due to their outstanding specific physical/mechanical properties, FRPs have attracted a great amount of interest from shipbuilding industry, offshore oil industry and marine infrastructure industry, used in recreational craft, workboats, risers for oil exploration and marine renewable energy device components. As for the carrier-based aircraft, the service life is about 25 years with an overhaul period of 10 years. When FRPs are exposed to marine environment (humid, thermal and saline) for such a long time, the degradation in mechanical properties has become the major issue for the safety and reliability of composites structures. Therefore, such mechanism of degradation serving in marine environment must be understood deeply to guide the engineering application. Li et al. [12] investigated the static/dynamic mechanical properties and failure mechanism of carbon fiber-reinforced polymer composites in seawater environments.

The prepared specimens were immersed in artificial seawater with different temperatures and NaCl concentrations for 7 months. Quasi-static tensile and DMA tests were carried out to evaluate the tensile and damping properties. Ageing damage and fracture morphology were observed by using SEM technology. The results show that tensile strength was sensitive to ageing time and environmental temperature but nonsensitive to NaCl concentration. In addition, the failure mode was changed. The damping value increased exponentially with the prolonged seawater ageing time. This was attributed to the fiber/matrix interface degradation of the composites. The higher the temperature was, the higher the damping value was, which indicated that the higher temperature caused more serious interface damage.

Many engineering structures made from CFRPs, including military equipment, automobiles, aircrafts, wind turbine blades and spacecrafts, often suffer from the menace of vibrations during their normal operations. Microcracks present in these structures propagate rapidly due to fatigue caused by vibrations, resulting in premature failure. Materials with high stiffness values generally have a low damping ability and so does CFRPs. Widespread applications of CFRPs in new areas are often limited due to their low damping factor resulting from poor viscoelastic nature of carbon fibers and poor damping at the CFRP interface. A few methods have been suggested to improve the damping properties of CFRPs, for example, addition of high damping polymer layers in prepreg lay-up and use of hybrid rubber particle reinforced composites. However, the improvements in damping properties arising from these methods are at the expense of reduced mechanical properties of the composites. With the development of CNTs and their composites in the last two decades, significant attention has been paid to improve the damping properties of polymers, especially epoxy, in terms of loss modulus or damping ratio. Khan et al. [13] studied the vibration damping characteristic of nanocomposites and carbon fiber-reinforced polymer composites CFRPs containing multiwall carbon nanotubes (MWCNTs) using the free and forced vibration tests. Several vibration parameters were varied to characterize the damping behavior in different amplitudes, natural frequencies and vibration modes. The damping ratio of the hybrid composites was enhanced with the addition of CNTs, which was attributed to sliding at the CNT–matrix interfaces. The damping ratio was dependent on the amplitude as a result of the random orientation of CNTs in the epoxy matrix. The natural frequency showed negligible influence on the damping properties. The forced vibration test indicated that the damping ratios of the CFRP composites increased with increasing

CNT content in both the 1st and 2nd vibration modes. The CNT–epoxy nanocomposites also showed similar increasing trends of damping ratio with CNT content, indicating the enhanced damping property of CFRPs arising mainly from the improved damping property of the modified matrix. The dynamic mechanical analysis further confirmed that the CNTs had a strong influence on the composites damping properties. Both the dynamic loss modulus and loss factor of the nanocomposites and the corresponding CFRPs showed consistent increase with the addition of CNTs, an indication of enhanced damping performance.

Material damping is a vital design component in the control of vibrations in various types of structural dynamics applications. A few benefits of incorporating materials with enhanced damping in structural design include: extending service life of components, attenuating acceleration and displacement responses, reducing noise, and possible reduction of total structural weight. In general, polymeric matrix composites (PMCs) provide higher material damping than metallic materials, where energy dissipation within fibrous composites may be attributed to: (1) viscoelastic/viscoplastic nature of the fiber and matrix constituents; (2) interaction between material phases; and/or (3) material damage in pre-existing or new structures. At the laminate level, the damping ratio of fibrous composites depends not only on the constituent lamina (or ply) properties, but also on ply orientations and interlaminar effects. Zhou et al. [14] designed a **c**arbon-fiber-reinforced hybrid-matrix composite (CHMC) to improve the material damping property in conventional carbon-fiber-reinforced epoxy (CF/E) through the interfacial reaction of two reactive polymeric compounds: still-curing epoxy-based phase I with a highly crosslinked morphology and lightly crosslinked still-curing polyurea elastomeric phase II. Chemical reactions of the hybridized interfacial matrix were discussed relative to phase reactions and tunable material damping in accordance with the migration of epoxy phase I species. Microstructures and micromechanical properties of CHMC were characterized using scanning electron microscopy and nanoindentation. Dynamic properties of CHMC were investigated via free vibration and forced vibration tests, and the results were compared to CF/E test results. Dynamic mechanical analysis revealed a cogent link between damping property (of the epoxy-polyurea interface) and lower $t_c \approx 0$, where proper design of the epoxy-polyurea interface could greatly improve material damping by using smaller epoxy curing times (t_c, hours) just before topically applying reactive polyurea. As a result, loss modulus of the interface was seven to ten times larger for $t_c = 0$ compared to polyurea-coated fully-cured epoxy. For

$t_c > 3.5$, the role of polyurea thickness h_p on energy dissipation expectedly increased.

From the above paragraphs, it could be inferred that damping is a critical parameter that defines the life of a structural component. The readers are advised to refer to recent research available on ScienceDirect for getting more information regarding the applications of damping in FRPs.

7.2 Principles of vibration control

Vibration damping methods are classified into passive and active methods. The control forces within the framework of an active control system are generated by a wide variety of actuators that have hydraulic, pneumatic, electromagnetic, piezoelectric or motor driven ball-screw actuation. In order to reduce the vibration of a helicopter, there are active rotor controls using piezoelectric materials that are intended to directly modify the periodic aerodynamic loads. By changing the geometry of a rotor blade, or by deflecting on-blade control surfaces, the aerodynamic loads can be modified and it can be considered to be passive vibration control. One of possible passive vibration control methods is usage of magnetic forces.

7.2.1 Active and passive vibration control for a thin-walled composite beam

Rimašauskienė et al. [15] developed and experimentally analyzed active and passive vibration control techniques for a thin-walled composite beam. An macro fiber composite (MFC) actuator was used, together with the magnetic forces between two permanent cuboidal magnets (passive vibration control) for vibration damping applications. The various lateral vibrations of the composite beam were caused artificially by the impact of an electro-magnet. The vibration control system that was developed consisted of forces caused by the MFC actuator (active vibration control) and the magnetic forces caused by two permanent magnets (passive vibration control). The first goal of this system was to find out what frequency (resonant or nonresonant) would be the most effective in achieving a higher vibration damping effect. The additional influence of magnetic forces on the damping of vibrations generated by the electro-magnet was also investigated. The experiments were carried out with and without the effect of magnetic forces, that is, changing the distance Δ between the permanent magnets and strengthening or weakening the influence of the magnetic fields and forces

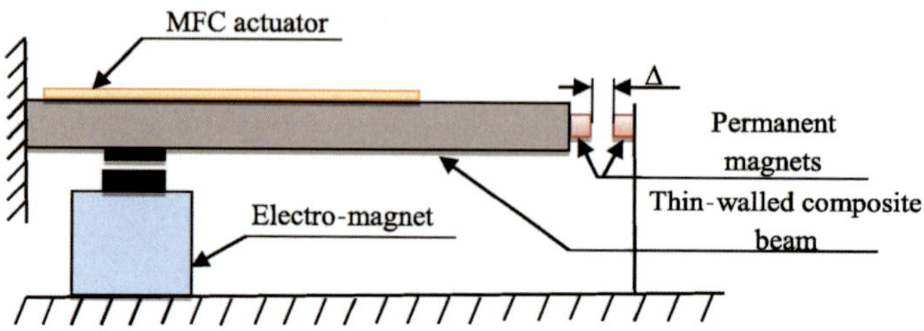

Figure 7.1 Design of the experimental vibration damping system (Rimašauskienė et al. [15]).

created between them (thus including them as passive vibration control). The vibration control system developed (Fig. 7.1) consisted of the composite beam (parameters 120 × 20 × 1 mm), two permanent cuboidal magnets (10 × 3 × 2 mm), an electro-magnet, and an MFC actuator M8514-P2 (active length 85 mm, width 14 mm, with an overall length and width of 100 mm × 18 mm).

To simulate and analyze passive and active vibration control in composite materials, a thin-walled beam was used. The permanent cuboidal magnets used in this system for passive vibration control imitation were made from a neodymium alloy. One of those magnets was glued onto the end of the thin-walled beam under analysis. Plus, another was used to have the possibility to change the distance, Δ, between the two magnets and was fixed on the micrometer screw. The vibrations in the thin-walled beam were introduced through the use of the electro-magnet. The MFC actuator (M8514-P2) (produced by Smart Material Corp.) (Fig. 7.2) was glued with an epoxy glue onto the surface of the composite beam 7 mm from the tightly fixed end.

After the experiments conducted with active vibration control (MFC actuator), it was found that in all cases, the vibration damping of the analyzed system was much higher at its resonant frequency. Nevertheless, due to the experimental results, it could be concluded that active and passive vibration control systems were efficient for thin-walled composite beam vibration amplitude control.

7.2.2. Passive control of flutter

The increasing use of composite materials in the aeronautical industry has made it possible to build lighter structures with better mechanical properties. Since lighter structures tend to be more flexible, undesirable vibrations and noise are increased and strategies that improve the aero-elastic stability

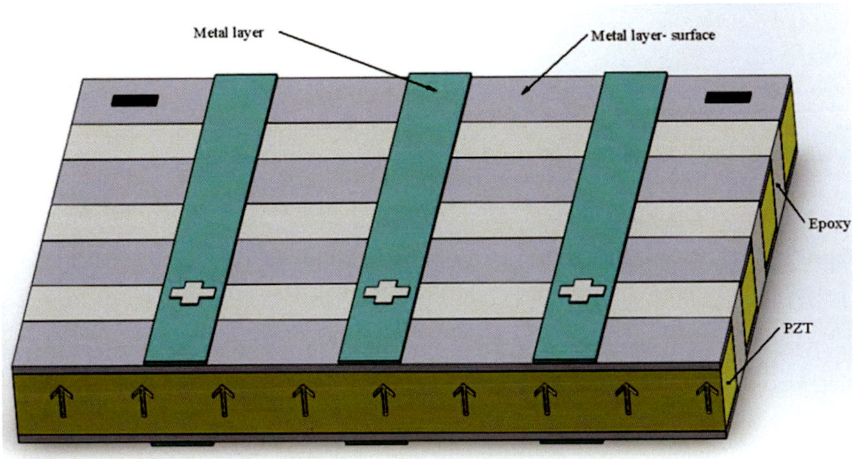

Figure 7.2 Structure of the macro fibre composite (MFC) actuator (type P2) (Rimašauskienė et al. [15]).

are key issues to be addressed. With the use of intelligent materials, it is possible that the structure has the ability to dissipate energy and stabilize itself when subjected to external dynamic loadings imposed by airflow. Ribeiro et al. [16] demonstrated the feasibility of using a passive control tool to increase the supersonic flutter boundary of composite panels using a multimodal shunted piezoceramic in parallel topology. For an application of industrial interest, uncertainties in the electrical parameters of the shunt circuit were considered in the control system design. Numerical applications were presented and the main features and capabilities of the proposed methodology were highlighted. Flutter, one of the instabilities, is caused by the interaction of elastic, aerodynamic, and inertial forces and produces large amplitudes of vibrations, modifying the way the aerodynamic loads act on a structure in an iterative and self-excited oscillatory process. Under this condition of instability, high internal stresses can occur to limit the vibration amplitudes of aeronautical panels that lead to structural fatigue. As a result, vibration and noise control strategies have been applied to aeroelastic systems to prevent flutter that can be catastrophic. Passive control techniques have been incorporated into many industrial systems because of advantages of low cost and easy maintenance and application compared with active control techniques. In addition, such techniques do not require an external energy source, which guarantees inherent stability of the system and makes them better adapted to aeroelastic systems. Typically, passive vibration control can be achieved using piezoelectric materials coupled to shunt

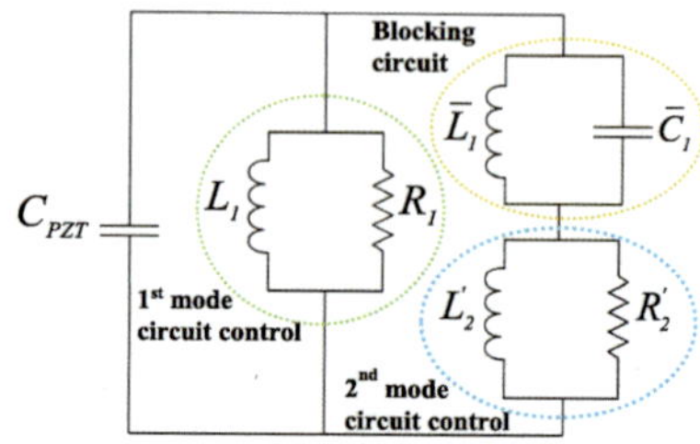

Figure 7.3 Resonant circuit in parallel topology for shunt damping of two modes.

electrical circuits. The multimodal circuit designed by Ribeiro et al. [16] was proposed by Wu (1998) with parallel topology. The circuit has the advantage of using a single piezoelectric patch to attenuate several structural modes simultaneously. The circuit shown in Fig. 7.3 was investigated for passive flutter suppression. The circuit was designed to simultaneously attenuate structural vibrations of the first two system modes. The first two modes coalesce when the composite structure was subjected to supersonic airflow. The proposed aeroelectromechanical system can help to define passive control strategies as a virtual testing tool for applications in composite flight vehicles. Furthermore, multimodal shunt circuits in parallel topology can be used as an alternative of a passive control technique to increase the flutter boundary structures; however, shunt circuit designer must carefully analyze the uncertainties in the circuit parameters as they can cause the system to behave differently than expected.

7.2.3 Active aeroelastic flutter analysis and vibration control

Song and Li [17] studied the active aeroelastic flutter analysis and vibration control at the flutter bounds of the supersonic composite laminated plates with the piezoelectric patches. As shown in Fig. 7.4, the piezoelectric patches were bonded on the top and bottom surfaces of the composite laminated plate to act as the sensor and actuator so that the active aeroelastic flutter suppression and vibration control for the supersonic laminated plate could be conducted. A four-ply supersonic composite laminated plate as shown in Fig. 7.4 was considered. Fig. 7.4 also shows the schematic diagrams of the laminated plate with piezoelectric actuator/sensor pairs.

The length, width and thickness per layer of the base composite laminated plate are a, b and h. The total thickness of the laminated plate is h_t. The piezoelectric actuator and sensor are bonded on the bottom and top surfaces of the plate, and positioned by the coordinates (x_1, y_1) and (x_2, y_2).

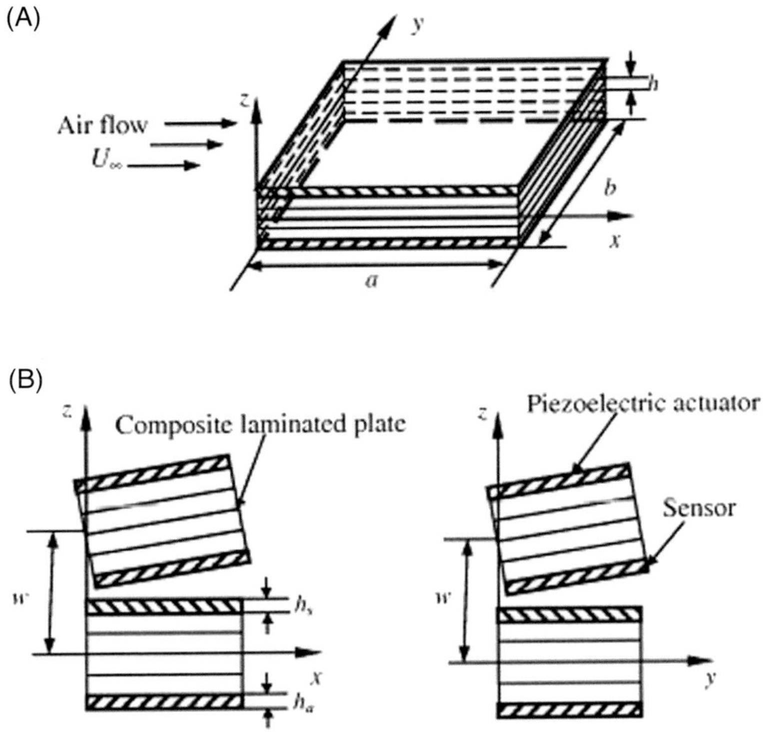

Figure 7.4 Schematic diagrams of the supersonic plate and piezoelectric actuator/sensor system: (A) the structural parameters of the structural system. (B) the kinematical parameters of the system.

The thicknesses of the piezoelectric actuator and sensor are h_a and h_s, respectively. Fig. 7.4B shows the schematic diagram of the displacements of the structural system. The unsteady aerodynamic pressure in supersonic flow was computed by using the supersonic piston theory. Hamilton's principle with the assumed mode method was used to develop the governing equation of the structural system. The controller was designed by the velocity feedback and proportional feedback control algorithm, and the active damping and stiffness were obtained. The solutions for the complex eigenvalue problem were obtained by using the generalized eigenvalue methodology. The natural frequencies and damping ratios were also evaluated. The aeroelastic flutter bounds of the supersonic composite laminated plate were calculated to investigate the characteristics of the aeroelastic flutter. The impulse responses of the structural system were calculated by using the Houbolt numerical algorithm to study the active aeroelastic vibration control. The influences of ply angle of the laminated plate and the control method on the characteristic of flutter and active vibration control were analyzed. From the numerical results it was observed that the aeroelastic flutter characteristics of the supersonic composite laminated plate could be improved and that the aeroelastic vibration response amplitudes could be reduced, especially at the flutter points, by the proportional feedback or the velocity feedback control algorithm using the piezoelectric actuator/sensor pairs. The effectiveness of the flutter control by the two control algorithms was also compared. The results of this study are of great significance to the flutter analysis and aeroelastic design of the aircraft.

7.2.4 Active vibration control of the laminated composite beams using velocity feedback control

The methodology for vibration suppression of structures can be categorized into two groups, namely passive and active controls. In passive control, the material properties of structure, such as damping and stiffness are modified so as to change the response of structure. A typical example of passive control is tailoring of composite materials. In active control, the structural response is controlled by adding external effort to the structure. Although active control requires the system to be complex and expensive, it can achieve a good vibration suppression performance in comparison with passive control. To sense and control structural response actively, a smart material, such as piezoceramic material, is frequently bonded onto the surface of a structure or embedded in it. Depending on the applied voltage, electromechanical coupling of the forcing transducer to the structure, and the location of the piezoelectric sensor/actuator, the degree of vibration control of flexible structures can be varied. Analytical and experimental studies on passive control by material tailoring and the active control by the use of piezoceramics have been studied separately, but the interaction between the two control approaches has not been investigated thoroughly. Kim et al. (2002) investigated the passive control by tailoring and active control by piezoceramic sensors/actuators numerically and verified experimentally. Structural vibration characteristics of smart composite structures were analyzed in order to study the dependence of the active damping on the feedback gain and the stiffness of the composite structures. Damping and stiffness of the adhesive and the piezoceramic layers were taken into account simultaneously in the finite element formulation. Experiments on the active vibration control of the beams were carried out by using the

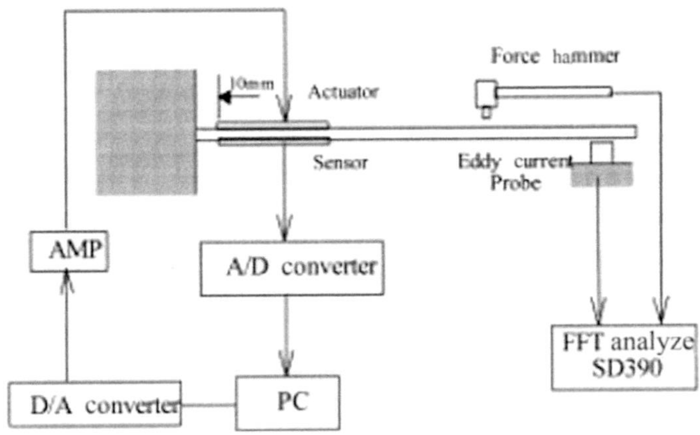

Figure 7.5 Experimental setup for vibration control of a laminated composite beam [19].

velocity feedback. Interaction between the passive and the active control was studied numerically and experimentally. Fig. 7.5 shows the model configuration and the experimental set-up for vibration control of the laminated composite beam with piezoceramic sensor/actuator. The dimensions of the piezoceramic in Fig. 7.5 are 50 × 20 × 0.5 mm. The adhesive material used to bond the piezoceramic sensor/actuator to the beam was cyanoacrylate adhesive. The measured thickness of the adhesive layer was 0.05 ± 0.01 mm. It was found that when the gain in velocity feedback control is small, the active control follows the trend of the passive control, but provides additional effects due to the active control. For a large feedback gain, the active control is dominant over the passive control. Active control is more effective in the structure with higher bending stiffness than in the structure with lower bending stiffness when the feedback gain is large.

7.2.5 Active vibration control of composite sandwich plate

Examples of practical active vibration control systems include active engine supports in vehicles and active vibration isolation systems for propeller aircraft. Depending on the problem, increasing stiffness and damping or isolating the structure are the most common solutions to obtain a vibration reduction. Stiffening the structure consists of shifting the resonance frequency beyond or over the frequency band of excitation. Increasing damping consists of reducing the resonance peaks by dissipating the vibration energy. Isolation consists of preventing the propagation of disturbances to sensitive parts of the systems. Several techniques are well known to increase the damping of a structure (passive vibration control) with fluid dampers, eddy currents, elastomers or hysteretic elements, or by transferring kinetic energy to dynamic vibration absorbers. Another common method is the use of transducers as energy converters that transform vibration energy into electrical energy that is dissipated in electrical networks, or stored (energy harvesting). Formerly, semi-active devices, also called semi-passive (passive devices with controllable properties) have been used. The magneto-rheological fluid damper and piezoelectric transducers with switched electrical networks are examples. Such techniques have very high rate of effectiveness, but when high performance is needed active control is the best choice. Active vibration control involves a set of sensors, a set of actuators, and a control algorithm; the design of the system implies many issues, such as the sensors and actuators configuration, and how to assure stability and robustness. The power requirement is crucial to define the size of the actuators and the cost of the setup.

Zippo et al. [18] tested an active vibration control of a free-edge rectangular sandwich plate. As a typical example, the bodywork and elements of the frame of racing cars and bikes can be mentioned. The experimental setup consisted of a honeycomb panel having a CFRP outer skins and a polymer-paper core, subjected to an orthogonal disturbance, due to an electrodynamics exciter and controlled by MFC actuators and sensors. MFC parches consist of rectangular piezoceramic rods sandwiched between layers of adhesive, electrodes and polyamide film. The MFC actuators and sensors were controlled by a programmable digital dSPACE® controller board. The control algorithm proposed in this research was based on the positive position feedback (PPF) technique and was successfully applied with different combinations of inputs/outputs (single input single output, MultiSISO, multi input multi output) in order to control the first four normal modes. The control appears to be robust and efficient in reducing vibration in linear (small amplitude) and nonlinear (large amplitude) vibrations regimes, although the structure under investigation exhibited a relativity high modal density, that is, four resonances in a range of about 100 Hz. The control strategy allows to effectively control each resonance both individually or simultaneously.

From the above paragraphs, the readers will be able to get an idea about how to control the vibrations in composite structures. The examples stated above are just a beginning, for details the readers are requested to refer to ScienceDirect (scientific papers are available) for getting more information about the principles of vibration control in composite materials. The authors have tried their best to include some of the significant techniques used nowadays for controlling the vibrations in composites. The readers will get introductory information about the principles of vibration control after reading this chapter.

References

[1] H. Hadiji, M. Assarar, W. Wajdi Zouari, F. Pierre, K. Behlouli, B. Zouari, R. Ayad, Damping analysis of nonwoven natural fibre-reinforced polypropylene composites used in automotive interior parts, Polym. Test. 89 (2020) 106692.
[2] Y. Li, S. Cai, H.Y. Liu, Y.W. Mai, Damping properties of carbon fiber reinforced composites hybridized with polysulfone (PSF)/cellulose nanocrystal (CNC) interleaves, Compos. Sci. Technol. 213 (2021) 108904.
[3] K.H. Khayat, W.J. Long, H.D. Li, L. Mei, W. Li, F. Xing, Damping characteristics of PVA fiber-reinforced cementitious composite containing high-volume fly ash under frequency-temperature coupling effects, Cem. Concr. Compos. 118 (2021) 103911.
[4] S. Liang, C. Zheng, F. Duan, Manufacturing and mechanical performance of novel epoxy resin matrix carbon fiber reinforced damping composites, Compos. Struct. 256 (2021) 113099.
[5] J. Zhao, H. Li, Z. Gao, H. Ma, Q. Han, J. Liu, Vibration suppression effect of porous graphene platelet coating on fiber reinforced polymer composite plate with viscoelastic damping boundary conditions resting on viscoelastic foundation, Eng. Struct. 237 (2021) 112167.
[6] Z. Sun, L. Qi, C. Li, X. Yu, W. Min, H. Shi, L. Tao, H. Wang, M. Yu, L. Ni, Effect of reinforced fibers on the vibration characteristics of fibers reinforced composite shaft tubes with metal flanges, Compos. Struct. (2021) 275.
[7] Y.A. Rao, K. Ramji, P.S. Rao, I. Srikanth, Effect of A-MWCNTs and etbn toughener on impact, compression and damping properties of carbon fiber reinforced epoxy composites, J. Mater. Res. Technol. 8 (1) (2019) 896–903.
[8] X. Wang, J. Zhang, A.A. Khatibi, E. Castanet, T. Baum, Z. Komeily-Nia, P. Vroman, Effect of natural fibre reinforcement on the sound and vibration damping properties of bio-composites compression moulded by nonwoven mats, Compos. Commun. 13 (2019) 12–17.
[9] G. Liaghat, A. Taherzadeh-Fard, A. Khodadadi, X.F. Yao, M.A.Z. Mehrizi, Mechanical properties and energy absorption capacity of chopped fiber reinforced natural rubber, Compos. Part C: Open Access 7 (2022) 100237.
[10] R. Pitchumani, C. DeValve, Experimental investigation of the damping enhancement in fiber-reinforced composites with carbon nanotubes, Carbon 63 (2013) 71–83.
[11] S. Kaewunruen, R. Janeliukstis, C. Ngamkhanong, Dynamic properties of fibre reinforced foamed urethane composites in wet and dry conditions, Mater. Today: Proc. 29 (1) (2020) 7–10.
[12] H. Li, K. Zhang, X. Fan, H. Cheng, G. Xu, H. Suo, Effect of seawater ageing with different temperatures and concentrations on static/dynamic mechanical properties of carbon fiber reinforced polymer composites, Compos. Part B: Eng. 173 (2019) 106910.
[13] S.U. Khan, C.Y. Li, N.A. Siddiqui, J.K. Kim, Vibration damping characteristics of carbon fiber-reinforced composites containing multi-walled carbon nanotubes, Compos. Sci. Technol. 71 (12) (2011) 1486–1494.
[14] H. Zhou, T.L. Attard, L. He, Improving damping property of carbon-fiber reinforced epoxy composite through novel hybrid epoxy-polyurea interfacial reaction, Compos. Part B: Eng. 164 (2019) 720–731.

[15] R. Rimašauskienė, V. Jūrėnas, M. Radzienski, M. Rimašauskas, W. Ostachowicz, Experimental analysis of active–passive vibration control on thin-walled composite beam, Compos. Struct. 223 (2019) 110975.
[16] L.P. Ribeiro, A.M. Gonçalves de Lima, Robust passive control methodology and aeroelastic behavior of composite panels with multimodal shunted piezoceramics in parallel, Compos. Struct. 262 (2021) 113348.
[17] Z.-.G. Song, F.-.M. Li, Active aeroelastic flutter analysis and vibration control of supersonic composite laminated plate, Compos. Struct. 94 (2) (2012) 702–713.
[18] A. Zippo, G. Ferrari, M. Amabili, M. Barbieri, F. Pellicano, Active vibration control of a composite sandwich plate, Compos. Struct. 128 (2015) 100–114.
[19] Y.K. Kang, H.C. Park, J. Kim, S.B. Choi, Interaction of active and passive vibration control of laminated composite beams with piezoceramic sensors/actuators, Mater. Des. 23 (3) (2002) 277–286.
[20] S.Y. Wu, Method for multiple-mode shunt damping of structural vibration using a single PZT transducer. In Smart Structures and Materials 1998: Passive Damping and Isolation (Vol. 3327, pp. 159-168). International Society for Optics and Photonics. 5th Annual International Symposium on Smart Structures and Materials, 1998, San Diego, CA, United States.

Index

Page numbers followed by "*f*" and "*t*" indicate, figures and tables respectively.

FSC
www.fsc.org
MIX
Papier | Fördert gute Waldnutzung
FSC® C083411

Kolibri 360 GmbH
Ferdinand-Jühlke-Straße 7
99095 Erfurt, Deutschland
produktsicherheit@kolibri360.de